User Interface Requirements for Medical Devices

User Interface Requirements for Medical Devices

Driving Toward Safe, Effective, and Satisfying Products by Specification

Michael Wiklund
Erin Davis
Alexandria Trombley

Illustrations by Jacqueline Edwards

CRC Press
Taylor & Francis Group
Boca Raton London New York

CRC Press is an imprint of the
Taylor & Francis Group, an **informa** business

First edition published 2022
by CRC Press
2 Park Square, Milton Park, Abingdon, Oxon, OX14 4RN

and by CRC Press
6000 Broken Sound Parkway NW, Suite 300, Boca Raton, FL 33487–2742

British Library Cataloguing-in-Publication Data
A catalogue record for this book is available from the British Library

---

Library of Congress Cataloging-in-Publication Data

Names: Wiklund, Michael, author. | Davis, Erin, author. | Trombley, Alexandria, author. | Edwards, Jacqueline, illustrator.
Title: User interface requirements for medical devices : driving toward safe, effective, and satisfying products by specification / Michael Wiklund, Erin Davis, Alexandria Trombley, Jacqueline Edwards ; illustrations by Jacqueline Edwards.
Description: Abingdon, Oxon ; Boca Raton, FL : CRC Press, 2022. | Includes bibliographical references and index. | Summary: "This book is a practical guide for individuals responsible for creating products that are safe, effective, usable, and satisfying in the hands of the intended users. The contents are intended to reduce the number of use errors involving medical devices that have led to injuries and deaths"—Provided by publisher.
Identifiers: LCCN 2021026072 (print) | LCCN 2021026073 (ebook) | ISBN 9780367457938 (hbk) | ISBN 9780367457471 (pbk) | ISBN 9781003029717 (ebk)
Subjects: LCSH: Medical instruments and apparatus—Standards. | Medical instruments and apparatus—Data processing.
Classification: LCC R856.6 .W55 2022 (print) | LCC R856.6 (ebook) | DDC 610.28/4—dc23
LC record available at https://lccn.loc.gov/2021026072
LC ebook record available at https://lccn.loc.gov/2021026073

ISBN: 978-0-367-45793-8 (hbk)
ISBN: 978-0-367-45747-1 (pbk)
ISBN: 978-1-003-02971-7 (ebk)

DOI: 10.1201/9781003029717

Typeset in Times
by Apex CoVantage, LLC

Contents

Acknowledgments

We are pleased to acknowledge several people and organizations who helped us produce what we hope will be a useful guide to specifying a medical product's user interface. We will start with a general salute to our consulting business customers.

For many years, we (three coauthors) have worked together in the Human Factors Research & Design consulting practice at UL (formerly titled Underwriters Laboratories and now delivering consulting services as Emergo by UL). Our work has included writing and reviewing user interface requirements for a wide array of medical product manufacturers based around the world. This experience highlighted for us the importance of user interface requirements in the drive toward great products. Accordingly, we are grateful to and thank our customers.

On a related note, we thank our consulting business colleagues from Concord, Massachusetts (USA); Chicago, Illinois (USA); Utrecht (The Netherlands); and Cambridge (UK). We have drawn many valuable lessons from our collaborations with them and our user interface design team members in particular (led by Cory Costantino).

Of course, we thank the acquisition and editorial team at Routledge, and Marc Gutierrez in particular, who accepted our book proposal in short order and helped us get to the publication finish line.

FIGURE 0.1   Applause.

We would be remiss if we did not offer thanks to the healthcare professionals at large, including physicians, nurses, technicians, therapists, maintainers, and many more role players. We have been part of a movement to make medical products safer through the application of human factors engineering processes and principles. Our work has called upon these medical professionals to share their vision of better products, provide feedback on evolving prototype designs, serve as usability test participants, and share their perspective on adverse events that have helped us drive product improvements.

And now, we will turn to personal acknowledgments.

MICHAEL WIKLUND'S ACKNOWLEDGMENTS

I thank my wife Amy Allen—an accomplished medical technology writer and editor—for enthusiastically supporting the book project and providing editorial support. Everyone needs a good emotional support system, and so I take this opportunity to acknowledge the support of my adult children, Ben, Ali, and Tom, as well.

I thank my colleagues Allison Strochlic and Jonathan Kendler with whom I started a private consulting firm (Wiklund R&D) that—via acquisition in 2012— evolved into the 70+ consulting human factors practice at UL. My professional path leading to prior writing projects and this book have depended heavily on our collaboration over the past 15 years.

Finally, I thank my colleagues (Dan Hannon, James Intriligator, Chris Rogers, and John Kreifeldt) at Tufts University, where I have taught human factors on a part-time basis since the late 1980s. They enabled me to build the Human Factors in Medical Devices and Systems certificate program and launch new, medical technology-focused courses within the human factors program at Tufts. Teaching these courses and interacting with many product developers and clinicians who have delivered guest lectures to the students have helped shape my views on medical product development and the needs of new practitioners.

ERIN DAVIS' ACKNOWLEDGMENTS

First, I thank my husband Chris Davis for his continuous support and encouragement of all my professional endeavors. Also, I thank my young children, Carson and Summer, who managed not to get into too much mischief during the hours I was focused on writing this book.

I thank my manager at UL, Allison Strochlic, who has given me so many opportunities to learn, grow, and be challenged as a human factors professional. Her encouragement makes me feel like I can tackle any task (including another book project).

Finally, I thank my professional family at UL. Their knowledge and experiences also informed the insights presented in this book. It is truly a joy to work among dozens of human factors specialists who share the same passion for thorough analysis, perfect grammar, and high-quality user interface design.

ALEXANDRIA TROMBLEY'S ACKNOWLEDGMENTS

First, I thank my fiancée Bradley Carlson, whose support energizes me during all of my professional and academic endeavors. You have always been a cheerleader for my career, faultless during late nights and demanding weeks, for which I will always be grateful. Also, I thank my family, including Whitney, Kirk, Gabriella, and my extended family, who have inspired me from a young age to speak up and given me the confidence to do so.

I also thank the community at my Alma Mater, Tufts University's School of Engineering. My advocacy for good human factors originated at Tufts, where I have the honor of continuing that advocacy as a part-time lecturer. In my professional and academic endeavors, I hope to inspire excitement about human factors and its impact, as Tufts inspired me.

Finally, I would like to thank my mentors and good friends at Emergo by UL HFR&D for encouraging me to pursue such professional ventures and equipping me with the tools to do so. Undoubtedly, many of the learning incorporated into this book reflect learning from our collaborations over the years.

JACQUELINE EDWARDS' ACKNOWLEDGMENTS

I thank my boyfriend, Bram Mulders, for his support and continuous supply of coffee (and the occasional cookie) during the long working days. You help me keep a calm, level head when I feel anxious and always know how to make me laugh. Also, thank you to my parents, Richard and Sandra, and my sister, Sarah. You all have cheered me on throughout my career and continue to inspire me to keep growing.

I thank the entire HFR&D team at UL. I have never been to a meeting that lacked laughter, and it is always a joy to work with all of you. I especially thank our Design Director and my mentor, Cory Costantino, for inspiring me to develop and challenge my design skills.

About This Book

This book is a practical guide for people responsible for creating products that are safe, effective, usable, and satisfying in the hands of the intended users. Part of the creative task of designing a medical product is to generate user interface requirements pertaining to its interactive elements and behaviors.

We—the authors—believe that a high-quality set of user interface requirements is central to producing a user-centered product and, subsequently, a successful product. Inferior requirements could spell doom for a product development effort because the requirements lead to a product that is misaligned with users' needs and preferences.

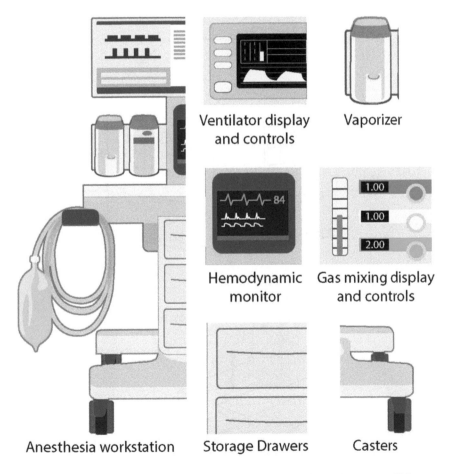

Ventilator display and controls

Vaporizer

Hemodynamic monitor

Gas mixing display and controls

Anesthesia workstation

Storage Drawers

Casters

FIGURE 0.2   Sample medical product (anesthesia workstation) "decomposed" into user interface subcomponents, each of which will warrant an extensive number of user interface requirements.

The construction of a sturdy house serves as a good analogy. To be sturdy, it needs to be built upon on a strong foundation, such as provided by a steel-reinforced footing and foundation wall (shown in blue in Figure 0.3). Absent of a strong foundation, there is no telling if the constructed home will stand up properly.

Today, in many regions in the world, there is a *de facto* requirement to develop and design products that meet preestablished user interface requirements. This is particularly the case with high-risk medical devices. The requirement has arisen at different times in different regions. It has been driven by the need to reduce the number of use errors involving medical devices that have led to injuries and deaths.

In brief, regulators expect manufacturers to take a structured approach to determining user needs and preferences and then meet them by converging on the right set of user interface requirements to create a product that will be resistant to use errors. This process is considered a dependable means to produce safe and effective medical products and one that is being ardently enforced by many regulators.

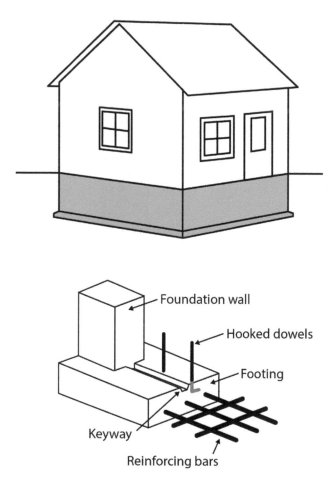

FIGURE 0.3   User interface requirements are analogous to a house foundation.

Let's consider an example of how good user interface requirements can lead to a safe and effective medical product. If you need a heart monitor to alert clinicians that a patient is having a heart attack, the monitor needs to draw their attention. It could do so by flashing an alarm message on the screen, emitting an audible alarm, sending urgent messages to a remote monitoring station, or all of these options. Each of these possible signals needs to be effective in view of the intended users and use environments, which means an audible alarm must have attention-getting characteristics (e.g., volume, frequency, burst).

In the United States, guidance published by the U.S. Food and Drug Administration is the principal driver for manufacturers to take a user-centered approach to medical product development, which includes developing user interface requirements. In the European Union, it is the Medical Device Regulation (MDR) and the International Electrotechnical Commission's (IEC's) standard on human factors (aka usability engineering) that are the driving forces. In many more countries, local regulators also enforce the IEC's standard on usability, which calls for user interface requirements to serve as the foundation for design. These and other standards and guidance documents ultimately call for manufacturers to set design expectations and meet them.

In FDA's parlance, manufacturers must create inputs (aka design requirements) and outputs (aka design features) and, once a product is complete, verify that there is a one-to-one correspondence and that there is a design feature and/or performance level to match each requirement. This expectation applies to a wide range of design features and not just to the user interface. For example, it pertains to the output of a power supply and the strength of a wheelchair's frame just as much as it does to the loudness of an alarm.

While the existing guidance documents and standards call for medical device design to be driven in part by user interface requirements, they are not sufficiently prescriptive about how to do it effectively. Also, medical product developers are

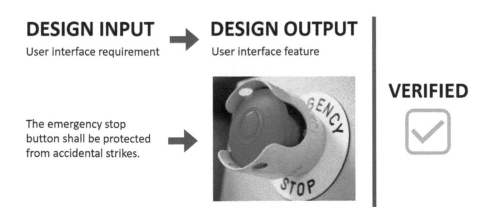

DESIGN INPUT → **DESIGN OUTPUT**
User interface requirement → User interface feature

The emergency stop button shall be protected from accidental strikes. →

VERIFIED ✓

FIGURE 0.4 Design verification is the process of confirming that design inputs match design outputs (i.e., that you designed the device right).

generally less experienced at developing comprehensive user interface requirements as compared to developing other types of requirements. This is because the intense application of human factors engineering in medical product design is relatively new.

This book can help people responsible for user interface design get on track. And, getting on track is imperative in view of the consequence of serious medical device use errors. It might sound simplistic, but you prevent user errors by design, and design is driven by requirements. Accordingly, you could say:

SAVE A LIFE! WRITE A SMART USER INTERFACE REQUIREMENT!

This book begins by affirming the strong connection between user interface requirements and safety. It then instructs readers how to develop specific requirements that are sufficiently comprehensive and detailed to produce good results—a user-friendly product that is likely to be used correctly.

The book's tutorial content is complemented by many examples of user interface requirements, including ones pertaining to an inhaler, automated external defibrillator (AED), electronic medical record, medical robot, and mobile app that a patient might use to manage her diabetes.

To ensure the book's readability, it adheres to these rules:

- Make your point
- Less is more
- A picture is worth a thousand words
- Stay focused on the primary topic

We hope that this book inspires and teaches readers to develop user interface requirements with a high degree of proficiency and that serve the interests of medical device developers, regulatory agencies, and, most importantly, the people who use medical devices. In fact, after digesting the content in this book and the aforementioned guidance and standards, readers may be well prepared to lead a user interface requirements development effort.

Content is grounded in human factors engineering science as well as on the practical knowledge of the authors. It includes multiple references to complementary literature, digital media, and organizations concerned with human factors engineering, use-related risk analysis, and overall medical device safety.

About the Authors and Illustrator

FIGURES 0.5–0.8 Michael Wiklund (top left), Erin Davis (top right), Alexandria Trombley (bottom left), and Jacqueline Edwards (bottom right).

Michael Wiklund serves as Director and General Manager of the Human Factors Research & Design practice, as part of UL's Enterprise and Advisory business. Michael is responsible for the direction and quality of the human factors consulting services that HFR&D provides to its clients. He identifies and develops new market opportunities where the consulting team can effectively apply its expertise at improving the relationship between people and technology, focusing specifically on use-safety, effectiveness, and usability. Historically, he has led projects requiring expertise in user interface research, design, prototyping, and usability testing. Michael joined UL in 2012 when the company he cofounded in 2005,

Wiklund Research & Design, was acquired by UL. He has authored eight books on human factors engineering and contributed to several national and international standards pertaining to human factors engineering. He is in his 35th year of teaching human factors at Tufts University, where he is Professor of the Practice. At Tufts, he helped established a certificate program for the study of human factors engineering as it applies to medical technology development. Michael is Corporate Fellow of the UL's William Henry Merrill Society. He is married to Amy and has three children and two grandchildren.

Erin Davis is an Associate Research Director with UL's Human Factors Research & Design team. She has been with the team since 2012. She has experience delivering HFE services to the medical device, pharmaceutical, scientific instrument, and laboratory equipment industries. A board-certified human factors professional, Erin leads and oversees research activities such as early-stage user research and usability testing. Furthermore, she helps medical device manufacturers develop key HFE documents for their design history files and advises clients on how to apply human factors engineering during product development to meet regulators' expectations. Erin is a coauthor of *Medical Device Use Error—Root Cause Analysis* and has served as President of the Human Factors and Ergonomics Society's New England Chapter. She holds a BS in biomedical engineering from Marquette University and an MS in human factors from Tufts University.

Alexandria Trombley is a Senior Human Factors Specialist with UL's Human Factors Research & Design team. She has been with the team since 2016. Alexandria leads a variety of human factors projects, ranging from early-stage user research to HF validation testing and HFE reporting. She collaborates with clients to develop key human factors deliverables, including user profiles and use environment descriptions, known problems analyses, risk analyses, and HF validation reports. Alexandria holds a BS in Human Factors Engineering and an MS in Human Factors Engineering, both from Tufts University, and is a part-time lecturer at Tufts University, teaching a class on Human Factors analytical methods.

Jacqueline Edwards is a User Interface Designer with UL's Human Factors Research & Design team. She has been with the team since 2019. Her design contributions aim to improve usability for a wide variety of medical products, from an injection device's instructions for use (IFU) to the graphical user interface for surgical equipment and everything in between. Jacqueline holds a BFA in Industrial Design from the Milwaukee Institute of Art and Design and an MS in Industrial Design from the Eindhoven University of Technology. She is also a part of the Diversity and Inclusion advisory team within the HFR&D team.

1 What Is a User Interface?

People interact (i.e., engage) with the world around them using their senses. Accordingly, when we interact with medical devices, such as a dialysis machine, syringe, infusion pump, or surgical stapler, we also use our senses. The parts of any such device that we see, hear, touch, smell, or taste comprise the *user interface*. Accordingly, the user interface might include hardware components, software screens, labeling (including user documentation), soundscapes, aromas (odors), material that literally has a taste, and more.

As illustrated in the following section, user interfaces may be one particular type or many (i.e., a hybrid).

HARDWARE

Here are some examples of hardware user interfaces or components therein, noting that the overall product might be a hybrid of hardware, software, and labeling.

Examples

Rotary knobs used to adjust the gas mixture (i.e., the amount of air, nitrous oxide, and oxygen)

Guardrail (bed rail) that keeps a patient from falling out of a hospital bed

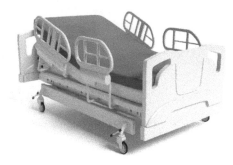

DOI: 10.1201/9781003029717-1

Examples

Package and blister pack containing vitamin capsules

Pushbutton ("M," for memory) used to recall and display the most recent measurement recorded by a glucose meter

Training-only electrode pads used to simulate delivering a cardioverting shock using an automated external defibrillator

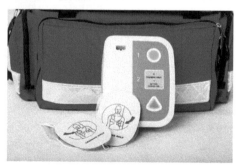

SOFTWARE

Even though a computer display is technically a hardware component, the display and the information presented on it (i.e., on screens) are normally referred to as the software user interface. Generally speaking, people use their eyes and a pointing device (e.g., mouse, fingertip, trackpad, stylus) to interact with information presented on a screen. Notably, a device's software user interface may incorporate more than one display.

Examples

Small, segmented LCD display that presents the temperature measurement taken by a forehead digital thermometer

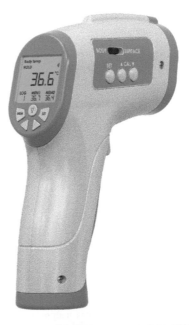

Tablet LCD display that presents hemodynamic parameter values and waveforms on a patient monitor

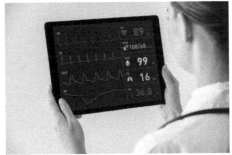

Examples

Large LED display that presents images of the coronary arteries, captured by an intravascular ultrasound imaging system

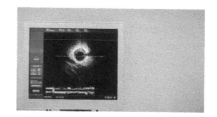

Touchscreen display, which displays information and touch controls, is built into a blood hematology analyzer

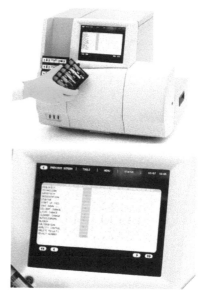

Smartphone and smartwatch screens display a patient's sinus rhythm data, which is analyzed to determine if there is atrial fibrillation

LABELING

Labeling is a term of art in the medical industry, referring to the words and symbols that might appear on hardware and software user interfaces as well as the various types of documents (virtual and printed) that accompany a medical device. Here are some examples.

Examples

Stop symbol printed on a syringe infusion pump's button

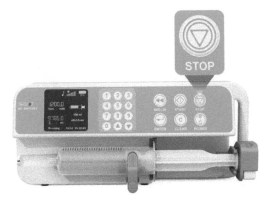

Enteral feeding product labels that enable clinicians to document the time and date

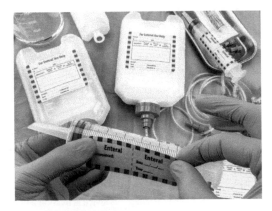

"Getting started guide" that introduces new users to a colon cancer screening kit

Examples

User manual that thoroughly describes a
tonometer's operation

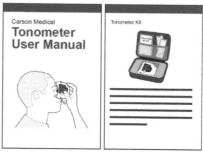

Package insert that provides step-by-step
guidance for performing a COVID-19 test

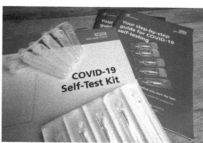

SOUNDSCAPES

The sound generated by a device, which might be produced by a speaker or contact between physical components, is also a user interface element. We refer to such stimuli as a *soundscape*. A soundscape may include the following types of sounds:

Examples

Alarms	Conditions: high heart rate, fluids line occlusion
Alerts	Conditions: calibration due, set dose level in a high range
Informative tones	Heartbeat, input confirmed
Clicks	Response to button clicks, knob rotation
Startup and shutdown melodies	Device activation and deactivation

AROMAS (ODORS)

A device might produce an aroma (aka odor) that influences how people interact with it. Here are some examples. (No, we do not use the term *aromascape* or *odorscape*.)

Examples

Aroma of insulin leaking from an infusion set adhered to a user's abdomen indicates a seal failure

Aroma of aerosolized drug emitted by a nebulizer confirms the nebulizer is working

Aroma of anesthetic agent escaping from an anesthesia machine's vaporizer during filling indicates a device failure

TASTES

You might be surprised, or perhaps even skeptical, that tastes can be part of medical device's user interface. However, tastes can play a key role in a user's interactions, especially with drug–device combination products that require the user to inhale or swallow medication.

For example, the aerosol produced by respiratory devices might coat the tongue on its way toward the trachea and lungs. In addition to any physical sensation that may result, the drug in an aerosol may produce a distinct taste that confirms that the device has delivered a dose. A particularly strong taste might indicate that too little drug remained airborne and will be swallowed rather than inhaled.

Taste also factors into the user experience when administering drugs via a medium designed to melt in the mouth, be chewed up and swallowed, or simply swallowed.

(No again. We do not use the term *tastescape* or *flavorscape*. ☺)

Tongue or teeth in the way of spacer/inhaler opening
FIX IT: Put the mouthpiece of the spacer/inhaler in the mouth above the tongue, under the top teeth.

FIGURE 1.1 Warning about misdirecting inhaled drug on the tongue rather than along the respiratory tract.

Source: National Jewish Health.

Examples

Taste of drug (inhaled dry powder) delivered by an inhaler confirms the drug's delivery

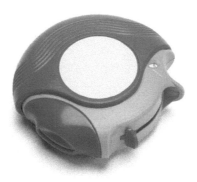

Taste of an aerosol emitted by a nebulizer confirms the nebulizer effectively nebulized the drug primarily for inhalation

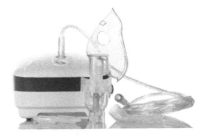

Examples

Taste of sublingual film (dissolvable strip placed
under the tongue) confirms drug delivery

INTEGRATED FORMS

Most of the more complicated medical devices include several, if not all, of the types
of user interface elements described earlier, including the following examples.

Examples

Anesthesia machine

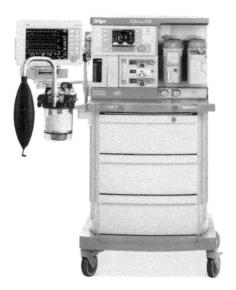

Examples

Hemodialysis machine

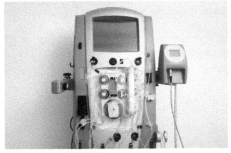

Hospital bed

C-arm X-ray machine

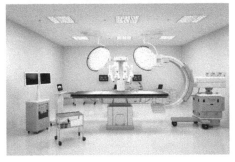

WHAT IS NOT A USER INTERFACE?

Given that a user interface is the part of a device with which people engage using their senses, the inverse is true. Portions of a device that people do not touch, see, hear, smell, or taste are not part of a device's user interface. This does not diminish the importance of these portions, which are usually crucial to a device serving its function, but they do not call for user interaction. Users might even be blissfully unaware of these portions' existence for that matter.

However, nonuser interface features and components can still indirectly impact the user interface and, accordingly, how users interact with the device. For example, if an internal power supply has failed and a device is running on backup battery power, this should trigger an audible and/or visual alarm via a related user interface element. Also, the power supply might be heavy and placed off-center within a device and accordingly influence how a user might grip the device during transport. In this case, the gripping surfaces are part of the user interface, but the user's experience is certainly influenced by the power supply's weight and placement.

Examples

Circuit board

Motor

Flow sensor

Spring

But, maybe we need to "back up the bus." What about the people who interact with such products during manufacturing, installation, and decommissioning? Such components certainly can be designed to facilitate such interactions to ensure safe, effective, and possibly even satisfying interactions. In fact, designing such components with human factors (HF) in mind can be quite important.

But (and that's two buts in a row), in this book, we limit ourselves to discussing what may be called the primary user interfaces of medical devices. That is, the elements related to a device's primary uses.

IF IN DOUBT . . .

By now, you probably have a good sense for what is and is not a user interface, at least as we have framed things for the purposes of this book. However, this should not limit the elements you consider worthy of inclusion.

The air filter that inserts into a surgical smoke evacuation machine strikes us as a good case of a user interface element that might be overlooked. A technician must replace the component on a routine basis. Accordingly, the component should be visually evident, clearly labeled, move easily in and out of its slot, and make it clear when it is properly installed in the machine.

The user will need to check the filter's label to be sure it is the correct one, orient it properly for insertion into its compartment, and insert it fully. One can imagine many potential use errors that could occur when interacting with the filter.

2 Role of User Interface Requirements in the Design Process

Think of user interface requirements as you might the ingredient list of a cake recipe. Keep in mind that the right ingredients in the right amounts are essential to baking a cake.

FIGURES 2.1 AND 2.2   Cakes and user interfaces both start with ingredients.

Now, what might happen if you bake a cake based on intuition and assumptions? You could mix a seemingly sensible amount of flour, sugar, water, eggs, and butter to start. Then, you could pour the mix into a pan and toss it into an oven set at a sensible temperature (let's say 250°F) for a sensible amount of time (let's go with 30 minutes). However, the result is likely to be disappointing. It is safe to say that this sad cake would be dense, undercooked, and not particularly flavorful.

Baking is a chemical process that requires precise portions of the right ingredients and proper cooking at the right temperature and for the right amount of time to produce a good result. And by the way, the cake we just cooked up was missing baking powder and salt, plus it should have stayed in the oven for 35–40 minutes at 350°F.

So, what is the parallel to cooking up a good user interface? Simple. It's all about the ingredients and how they are "cooked" (i.e., designed).

DOI: 10.1201/9781003029717-2

Let's now consider the creation of pen-injector.

FIGURE 2.3   Insulin pen-injectors.

Below, we present an "ingredient" list:

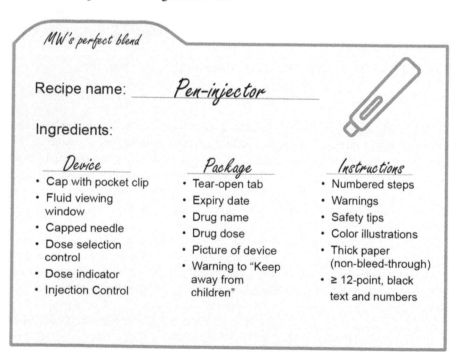

FIGURE 2.4   Ingredients for a pen-injector's design.

Similar to a cake, a user interface assembled strictly by experience and intuition might be disappointing in the sense that it is not to the user's taste, so to speak. In other words, the user interface might not meet the user's needs or preferences to the positive extent that it could if designed according to a recipe.

User interface requirements are essentially an ingredients list for a user interface. The blending and cooking of the ingredients are, metaphorically speaking, the design process.

Logically, you should establish at least some of the guiding user interface requirements at the early stage of a product development project, perhaps as the first step ahead of forming a design vision or provisional design concepts. This approach resonates with the designers' mantra that form should follow function. We would like to extend this mantra by stating:

FORM FOLLOWS FUNCTION FOLLOWS USER NEED

Basic user interface requirements can be derived from the following sources:

- Requirements (or entire specifications) defined for predecessor devices, noting that these requirements might be dated or off-target in some way and subject to replacement later on as new information becomes available.
- Requirements derived from design guideline content presented in various sources (e.g., AAMI HE75,[1] Apple's Human Interface Guidelines,[2] articles or blogs focused on the type of product being developed; see Chapter 12 for additional resources and details).
- Knowledge about available technologies, such as the necessary resolution of a display.

Subsequently, many more and much more specific user interface requirements should arise from a series of research, analysis, design, and evaluation tasks. This is the advantage of a user-centered design process. Whereas there is no practical means to change the ingredients of a cake once it is baking, you can change the ingredients of a user interface. You can add or subtract display content, controls, and labels, for example. Also, you can rearrange user interface components to best suit the users' workflow and specific tasks. So, this is where we will say goodbye to the cake metaphor. User interface requirements can be modified until almost the end of the design process at which point you will validate the final design and then transfer it to manufacturing.

During the design process, the many sources of user interface requirements and causes for "tuning" them come from activities such as the following major ones:

- User research that identifies users' needs and preferences
- Hazard, task, and risk analyses that identify ways to mitigate risk through user interface design solutions
- User interface design efforts that may explore multiple design solutions and reveal the advantages of a particular one

- Formative evaluations, including usability testing, that lead to insights on how to improve the user interface
  - Changes to other aspects of the design (e.g., electronics, battery, material) that impact the user interface

As such, an initial set of user interface requirements may be reverse-engineered as a result of subsequent design process steps, but this is natural and desirable. There is no point in sticking with an initial requirement once you gain insights on how to make it a better one, just as long as the adjustment does not undermine an original, legitimate, and important intent.

DOCUMENTING THE RATIONALE FOR USER INTERFACE REQUIREMENTS

We see considerable value in documenting the rationale for each user interface requirement. The rationale should include the requirement's source(s), such as a user need, predecessor product's documentation, or a design principle. If a requirement is revised over time, it is also useful to document the rationale for the changes.

Although writing the rationale adds more work initially, it is something you will thank yourself for later. The benefit of this added effort is that it establishes a history of the user interface requirements, thereby increasing corporate memory of why the requirements were written in the first place. In the absence of having documented rationales, newcomers to the team or those less familiar with the user interface design might remove or revise user interface requirements in a way that compromises user needs and leads to poor and potentially unsafe user interactions.

A secondary benefit of including user interface requirement rationales is that it can make future requirement development efforts more efficient. Let's say you are kicking off the development of a new product and want to leverage some user interface requirements from the predecessor product. If the predecessor's user interface specification includes a rationale for each requirement, it's easier and faster to recognize which requirements you can leverage for the new product and which you can disregard. For example, user interface requirements related to design principles might still be relevant (and conveniently you won't have to look up the principle's reference *again*), while you can ignore requirements related to any obsolete user needs that are not applicable to the new product.

The following table tracks how an initial user interface requirement for a magnetic resonance imaging (MRI) machine can evolve into multiple requirements over the course of a product development effort. Note that the bold text indicates the modification(s) at various stages.

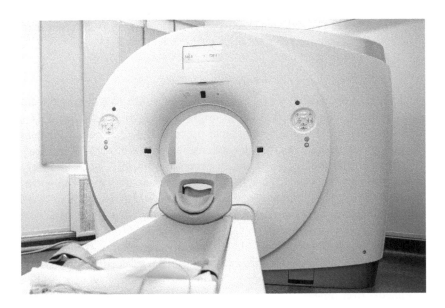

FIGURE 2.5   An MRI machine, which includes two red emergency stop buttons.

Source	User Interface Requirement(s)
Predecessor product's requirement	• **There shall be an emergency stop button (E-stop).**
Interviews with prospective users	• There shall be an emergency stop button (E-stop). • **The E-stop shall be colored red and be surrounded by a yellow collar.**
Hazard analysis	• There shall be an emergency stop button (E-stop). • The E-stop shall be colored red and be surrounded by a yellow collar. • **The E-stop shall be protected from accidental actuations by a clear cover.**
Use-related risk analysis	• There shall be **two** emergency stop buttons (E-stops). • **Each E-stop** shall be colored red and be surrounded by a yellow collar. • **Each E-stop** shall be protected from accidental strikes. • **One E-stop shall be placed on the movable gantry.** • **One E-stop shall be placed on the remote control panel.**
Conceptual design	• There shall be two, **identical** emergency stop buttons (E-stops). • Each E-stop shall be colored red and be surrounded by a yellow collar. • Each E-stop shall be protected from accidental strikes. • One E-stop shall be placed on the movable gantry • One E-stop shall be placed on the remote control panel.
Formative usability testing	• There shall be **three,** identical emergency stop buttons (E-stops). • Each E-stop shall be colored red and be surrounded by a yellow collar. • Each E-stop shall be protected from accidental strikes. • **One E-stop shall be placed on the magnet's left-side control panel.** • **One E-stop shall be placed on the magnet's right-side control panel.** • One E-stop shall be placed on the remote control panel. • **The E-stops shall be labeled "Emergency Stop."**

The various stages of the design process and how they serve to shape the evolving set of user interface requirements are discussed in Chapters 7–11.

A Living Document

People describe a document containing risk analysis results as a "living document" because it is never quite finished. Analysts might create an initial risk analysis based on what is known early in the development process. But, it is bound to change—mostly expand—as more is learned during ensuing stages of the development process. Even when the product development process is finished and a product goes to market, the risk analysis still continues based on post-market surveillance data among other things. This is actually required by *ISO 14971:2019, Medical devices—Application of risk management to medical devices*, which states that "risk management does not stop when a medical device goes into production."[3]

Manufacturers should also apply the "living document" model to a set of user interface requirements in response to newly identified user needs, known problems, and risks. In other words, the document should evolve—mostly expand and become more refined—over time.

Of course, this implies the need to plan for intermittent updates and for careful version control. The following table indicates points during a development process when the starting set of user interface requirements is likely to change, presuming that an initial set will be compiled ahead of use-related risk analysis, which is not always the case. For illustration purposes, we have assumed that we are developing a blood glucose meter (i.e., glucometer) used by people who have diabetes to check their blood glucose (i.e., sugar) levels.

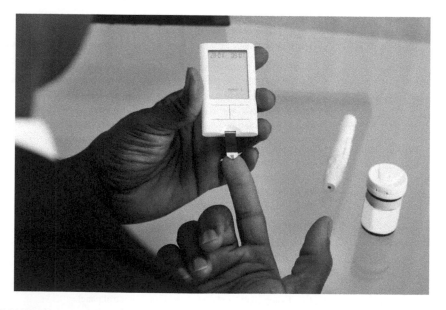

FIGURE 2.6   A person with diabetes touches a blood droplet to a blood glucose meter's test strip.

Development Phase/Activity	Sample Changes	Resulting User Interface Requirement
During user research	Interviews with prospective users reveal that users sometimes struggle to interpret numerical expiration dates (e.g., 01–10–22) and that they prefer to see the month indicated using letters (e.g., "NOV" for November rather than "11").	The test strip expiration date month shall be presented using letters.
During use-related risk analysis (including "feeder" activities)	Known problems analysis suggests users often do not notice and/or confirm the blood glucose units of measure.	During device setup, the device shall require users to confirm the units of measure in which blood glucose readings are presented.
	Hazard analysis suggests adding a warning to the device, package, and instructions for use to keep the device dry (i.e., do not splash with water or submerge in water).	Device shall alert user to the potential harm associated with the blood glucose meter getting wet.
	Task analysis suggests the steps to power-on the meter and insert the test strip might be combined to reduce the overall number of steps.	The device shall power on automatically when it detects the insertion of a test strip.
During preliminary design	User interface designers determine it is not feasible to incorporate a high-fidelity speaker into the meter, and so the "soundscape" requirements must be modified to match a low-fidelity, piezoelectric emitter.	Revise existing requirements to reflect technical specifications of available speaker.
After early formative usability tests	Early formative usability testing suggests shaping the meter so that takes little hand strength to grip it securely.	The device shall fit within a fifth percentile female's palm.
During detailed design	User interface designers recognize the opportunity to add a light at the test strip insertion point.	Test strip insertion point should be illuminated.
After late-stage formative usability tests	Late-stage formative usability testing suggests coloring the control solution (used to confirm the meter is working properly) so that it is visible on a test strip.	The control solution shall be colored rather than clear/opaque white.
During final design	User interface designers recognize the need to reduce screen glare.	The screen's luminance ratio shall be at least 3.[4]
After a "failed" HF validation test that included root cause analysis of observed interaction problems	A "failed" HF validation test (i.e., another formative usability test) suggests displaying the blood glucose measurement (e.g., 125 mg/dL) substantially longer than just 5 seconds.	The blood glucose reading shall remain on the blood glucose meter's screen until the device powers off or another blood glucose reading is presented.
Over the course of post-market surveillance	Based on reports of devices running out of battery power quickly, the manufacturer recognizes the need for device to turn off automatically after an appropriate period of time.	The blood glucose meter shall power off after 60 seconds of inactivity.

PUTTING USER INTERFACE REQUIREMENTS TO THE TEST DURING FORMATIVE USABILITY TESTING

The purpose of **form**ative usability tests is to "test drive" the product while it's still developing (i.e., while it's still "**form**ing"). Formative usability tests are valuable because they reveal a product's interactive strengths and—perhaps most importantly—a product's opportunities for improvement. Often, formative usability tests uncover issues with the design that warrant design changes, but the underlying user interface requirement is sound. In other cases, the user interface requirements might actually need some modification to ensure they align with user needs and can serve as a solid foundation for design activities.

Let's look at an example series of events that demonstrate how a formative usability test can uncover a flawed user interface requirement.

1. Early in development, user research indicated that users feel overwhelmed when presented with too much information at once while reading IFU.
2. This research finding led to a user interface requirement, which stated that "the instructions for use shall present no more than two steps per page," thereby limiting the amount of information a user will see at one time.
3. This requirement ultimately led the design team to develop a booklet-style IFU.
4. During several formative usability test sessions, the test team observed participants struggling to concurrently flip through booklet pages, hold the meter, and keep their lancet-pricked (i.e., bloodied) finger from touching anything. In fact, several debriefed participants recommended getting rid of the booklet altogether in favor of a format that would not require them to do any paging.
5. After reflecting on the formative usability test findings, the team concluded that the initial user interface requirement was overly specific and that users might be better served by format-agnostic user interface requirements. The team developed more general user interface requirements that aligned with design principles for labeling (e.g., including numbered steps, using headers to convey a clear hierarchy, avoiding high information density by maintaining a reasonable amount of white space).
6. During subsequent formative usability tests, a "fresh sample" of participants gave high compliments to the newly designed large, single-page IFU. Participants appreciated that they did not have to hold the instructions while interacting with the meter. They also considered the instructions inviting and easy to follow.
7. After the test, the team celebrated their design victory during a lovely group dinner.

NOTES

1 Association for the Advancement of Medical Instrumentation. 2009. *ANSI/AAMI HE75–2009: Human Factors Engineering—Design of Medical Devices*. Arlington, VA: Association for the Advancement of Medical Instrumentation.
2 Apple Developer. n.d. *Human Interface Guidelines*. https://developer.apple.com/design/human-interface-guidelines
3 International Organization for Standardization. 2019. *14971:2019 Medical Devices—Application of Risk Management to Medical Devices*. Geneva, Switzerland: International Organization for Standardization. www.iso.org
4 Association for the Advancement of Medical Instrumentation. 2009. *ANSI/AAMI HE75–2009: Human Factors Engineering—Design of Medical Devices*. Arlington, VA: Association for the Advancement of Medical Instrumentation.

3 Why We Need User Interface Requirements

SATISFYING REGULATORY REQUIREMENTS

Since 1996, the Quality System Regulation (QSR) in the United States has called for manufacturers to develop products in a more user-centered manner. *Subpart C—Design Controls, § 820.30 Design controls, Subpart (c) Design input* of the regulation states:

> Each manufacturer shall establish and maintain procedures to ensure that the design requirements relating to a device are appropriate and address the intended use of the device, including the needs of the users and patient.

So, there you have it. A clear signal from one of the world's leading regulatory bodies that a medical device developer needs to establish user interface requirements linked to users' needs and—we recommend—users' preferences.

Establishing user interface requirements not only helps guide the ensuing product development effort but also provides a basis for satisfying another QSR requirements *Subpart (f) Design verification*, which states:

> Each manufacturer shall establish and maintain procedures for verifying the design input. Design verification shall confirm that the design output meets the design input requirements.

Note that the FDA's Design Controls process uses the term "design inputs" as a synonym for "requirement" and the term "design outputs" as a synonym for "feature." Whichever pair of terms you use (requirement/feature or design input/design output), verification amounts to a matching game and is a relatively straightforward, analytical exercise.

Following design verification, manufactures are then obligated to ensure that their final product can actually be operated in a safe and effective manner by the intended users. This requirement is captured in *Subpart (g) Design validation*, which states:

> Design validation shall ensure that devices conform to defined user needs and intended uses, and shall include testing of production units under actual or simulated use conditions.

As compared to verification, validation is usually a much more energetic exercise involving representative users ostensibly taking a product for a "test drive" and encountering critical situations.

DOI: 10.1201/9781003029717-3

23

To illustrate the workflow, consider the tasks of identifying a user need, establishing a requirement, designing a user interface that meets the requirement, and then verifying and validating that you've met the requirement and the user need, respectively, pertaining to an AED.

USER NEED

Users need to be able to rapidly recognize and press the shock button.

USER INTERFACE REQUIREMENT

There shall be a single-purpose shock button.

USER INTERFACE DESIGN

FIGURES 3.1   Sample AED and close-up of shock button.

VERIFICATION

Inspect the AED to confirm that there is a single-purpose shock button (i.e., confirm that there is a feature that matches the requirement). Document that the feature exists in a verification report.

VALIDATION

Conduct a usability test of the AED to confirm that the intended users are able to recognize and operate the single-purpose shock button safely and effectively when engaged in realistic use scenarios. Document the safe and effective performance of the single-purpose shock button-related use scenario(s) in a report. In other words, document that the intended users were able to deliver a shock to a simulated patient in a representative use environment.

Now that we have covered the regulatory requirement in the United States, what about other territories? Do you have to develop user interface requirements for products that might be sold in the European Union (EU), for example? The answer is yes.

The EU and several other territories call for medical devices to comply with a broad-based standard titled *IEC 62366–1:2015 Medical devices—Part 1: Application of usability engineering to medical devices. Section 5.6 Establish User Interface Specification* states:

> The User Interface Specification shall include . . . testable technical requirements relevant to the user interface, including the requirements for those parts of the user interface associated with the selected risk control measures.

While it is more easily understood after reading the standard in its entirety, IEC 62366–1:2015 essentially requires manufacturers to establish user interface requirements, thereby matching the FDA's expectations set years earlier. The standard allows manufacturers to focus on a user interface's safety-related aspects as opposed to the user interface in totality. However, in practice, manufacturers are well served to establish requirements for the entire user interface because a comprehensive set of requirements is needed to control the design process and achieve the desired outcome.

SATISFYING COMMERCIAL OBJECTIVES

Another reason to establish user interface requirements is to win in the marketplace by producing a product that delivers a great user experience. Remember that designer talent and luck will only occasionally bring about this outcome. More often, a commercially successful product arises by understanding users, use environments, and use scenarios and then generating appropriate user interface requirements. And, the key word is *appropriate*.

Imagine the user interface requirements that could be associated with each of the following chairs.

Each one of these chairs may be considered a grand success or failure depending on the design intent, which we can relate directly to the underlying design (user) requirements. The intent might have been to maximize restfulness, optimize

FIGURES 3.2–3.4 The designers of these three chairs likely had very different design objectives.

ergonomics, or make a design statement. No doubt that the intent was also to produce a commercially successful product, targeted toward a particular market segment. Regarding the showroom of chairs presented earlier (moving left to right), the targets were very likely the following:

- People who love "coziness" and perhaps like to watch football on TV and possibly fall asleep in place.
- People who value excellent ergonomics while performing deskwork and might also want to sit in chair that fits in a workplace with a high-tech milieu.
- People who are drawn to a stylish, brightly colored design.

The notion of targeting certain consumers by envisioning a specific solution and developing appropriate user interface requirements is not so different for medical devices. Here is a sample of wheelchairs that serve different types of users, albeit while serving the same, basic functional requirement of getting people where they want to go.

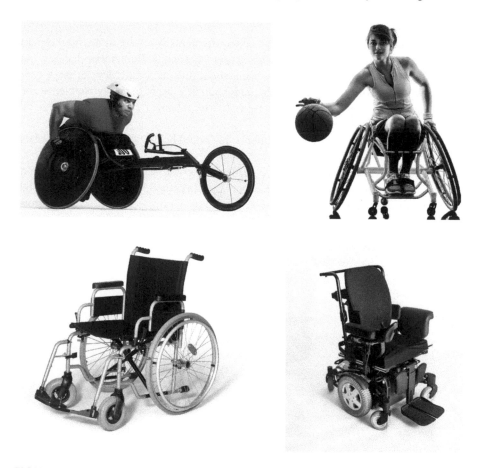

FIGURES 3.5–3.8   Four wheelchairs that address different market needs.

Clearly, producing a commercially successful wheelchair will depend on determining and meeting the right requirements. For example, in one case, the ability to speed along a road in a lightweight product (top left) is paramount, and in another case (top right), quick maneuverability might be crucial during particular athletic endeavors. In a third case (bottom left), the ability to fold the wheelchair is the dominant capability, whereas in a fourth case (bottom right), rock solid stability and electric powered movement might be essential to meeting users' needs.

Note that the user interface requirements associated with the wheelchairs likely have overlap with what you might consider to be basic functional requirements or perhaps even electromechanical requirements. As far as we are concerned, this overlap is just fine.

ADDING STRUCTURE TO THE DESIGN PROCESS

Another advantage to writing user interface requirements is that it helps structure the ensuing design efforts. Instead of having a designer or team of designers simply intuiting a solution, she, he, or they can review user interface requirements to create a base understanding upon which to build while still drawing on their intuition, which really means drawing on creativity and past experience.

The structure added to the design process by establishing user interface requirements is a good way to avoid . . . let's call it *creative chaos*. This is not to say that some "free styling" or "creative adventures" are not a good thing. Designers might want to indulge in an unlimited creative exercise at the very beginning of the design process just in case it reveals some breakthrough ideas. But then, the application of some structure is a wise next step. Otherwise, you can end up with some wild, blue sky, and super creative design solutions that simply don't work; that are not aligned with users' actual needs and preferences and, therefore, are a bit wasteful.

4 Common Pitfalls When Writing User Interface Requirements

The Merriam-Webster online dictionary defines a pitfall as "a hidden or not easily recognized danger or difficulty."[1] So, what possible difficulty or danger could someone writing user interface requirements encounter? Very little if you take this question literally. Writing user interface requirements is a pretty benign activity if you use an ergonomic keyboard or do not poke your eye with a pen. But taking a less literal stance, the nuanced process of writing user interface requirements presents several difficulties that could—if not addressed—lead to sub-par requirements. We highlight some of these pitfalls in the following sections.

FIGURE 4.1   Watch out for pitfalls.

DOI: 10.1201/9781003029717-4

LACK OF DATA

Suppose that you are concerned about making a portable medical device sufficiently lightweight. You could draft the following user interface requirement:

- The device's weight shall not exceed X lb.

But, what is the right value for X?

A laptop weighs about 2.8 lb. An intravenous infusion pump weighs 2.5 lb. A point-of-care ultrasound machine weighs 3.1 lb. So, can we assume that a weight limit of 3.0 lb is a good one? It might or might not be. You have strong evidence that 3 lb (or so) is a common weight for some portable devices; however, you still do not know that a 3 lb version of the device you are developing would be acceptable to the device's users. Therefore, you might consider conducting some user research with the intended user population to generate the data necessary to write a good requirement.

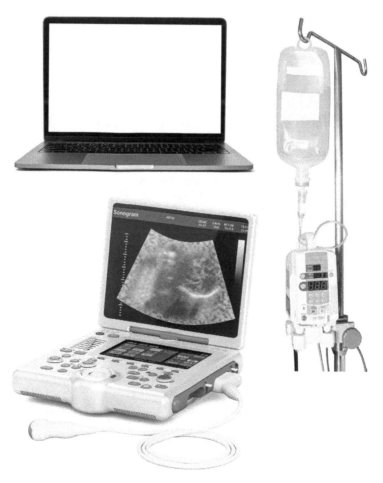

FIGURES 4.2–4.4   Laptop, infusion pump, and portable ultrasound machine.

NO MEANS OF ENFORCEMENT

In principle, this pitfall should not present itself in the course of developing a medical product. A manufacturer's quality management system should ensure that all user interface requirements are met, obviating the possibility that the development team would ignore some of them. However, the reality is that requirements can be overlooked or even intentionally disregarded, particularly if the requirements present development or manufacturing challenges or feel like a bureaucratic obstacle. If discovered late in the development process, user interface requirements might even be deleted as a matter of convenience and time savings. Historically, it has been too easy for engineering teams lacking a full appreciation of user needs and preferences to dismiss user interface requirements. Such dismissals might compromise a product's safe, effective, and satisfying use. Therefore, one facet of producing user interface requirements that drive user interface quality and safety is ensuring that quality management processes are in place and—to be blunt about it—enforced.

Manufacturers striving to ensure their team consistently applies user interface requirements might consider implementing a requirements program, resulting in a central repository of requirements and status information. Project managers can leverage the repository to ensure the project and product are on track, and the repository can be applied to future product generations.

MUTUAL EXCLUSIVITY

It is common for several individuals, groups, or even organizations to contribute to a comprehensive set of user interface requirements. That said, if the contributors are siloed, without awareness of the requirements developed by their colleagues, they might generate mutually exclusive pairs (or sets) or requirements. Here are several pairs of opposing requirements:

- The handle shall be textured to facilitate a secure grip.
- The handle shall be smooth to facilitate cleaning.

- Audible alarm volume shall be adjustable over the range of 60–80 dB.
- High-priority audible alarms shall annunciate at 80 dB.

- The color red shall be used exclusively to indicate high-priority alarms.
- Blood pressure values shall be colored red.

This poses obvious problems because there is no way to address both requirements in each pair.

Be sure to look for these discrepancies (or ask a detail-oriented colleague to do so) and ensure you correct the issues before passing user interface requirements onto designers for implementation. In most cases, the conflict can be resolved by deleting the less important requirement, refining the requirement(s) to be more precise, or adding clarifying details. These resolutions are reflected in the following, modified requirements:

- The handle shall be lightly textured to facilitate a secure grip and enable effective cleaning.

- Low- and medium-priority alarms shall be adjustable over the range of 60–80 dB.
- All high-priority alarms shall annunciate at 80 dB.

- High-priority alarms shall be displayed with white text on a red background.
- Blood pressure values shall be colored red.

INCOMPLETE

The user interface requirement cited earlier (Critical controls shall be protected from accidental strikes) provides good "coverage" by including the term "critical controls" and identifying the controls that are considered critical. The requirement would be incomplete if there were multiple, critical controls, but the requirement was limited to an emergency stop button and excluded other critical controls, such as the power switch.

TOO SPECIFIC

To ensure your user interface requirements are complete and can be verified, you might be inclined to make them very specific. Inherently that is not problematic, unless the requirement crosses the line between specific and prescriptive. Let's take a look at the user interface requirement later:

- A text label shall be placed above each physical control.

While well intentioned, this requirement precludes potentially good design solutions, such as labeling a power button with the conventional symbol for power on/off and placing the label on the button itself.

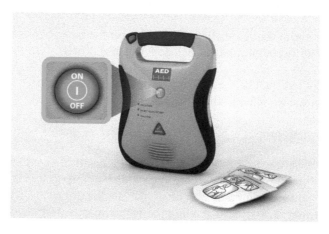

FIGURES 4.5   AED's power button.

Seek opportunities to present complete, comprehensive user interface requirements that can be verified, without prescribing the user interface design elements that then address those requirements.

That said, there are some situations when being more specific is necessary (and valuable). We suggest writing user interface requirements that are more prescriptive when there is a mandate for a specific design solution, which might be found in an industry or corporate standard. Here is a sample of highly prescriptive requirements that essentially rule out other design options. Note that they deliberately leave little, if any, room for consideration of other options.

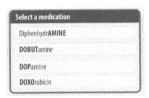

Drug names shall be written using TALLman letters.[2]

The CPR button shall be colored red.

The power switch shall be located on the front of the device.

All alarm symbols shall conform with those shown in IEC 60601–1–8. Also, see Chapter 13—Alarms.

Systolic and diastolic blood pressures shall be displayed in red alphanumerical characters.

These requirements could be "softened" to be less prescriptive, but the requirements might open the door to less effective design solutions and depart from accepted, standard practice.

COMPOUND REQUIREMENTS

It's smart to give each user interface requirement a singular focus. Conversely, it is . . . not so smart . . . to write requirements addressing more than one point. The primary reason is that a compound requirement complicates the design verification process, one that is well served by binary outcomes whereby a requirement is either "met" or "unmet." "Partially met" has no place in the verification vocabulary.

Writing requirements that address one point at a time might seem inefficient and even tedious at times. However, it pays off down the line when the time comes to verify a design, answering the question, "Does the design meet the requirement or not."

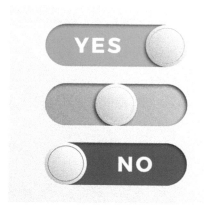

Later, we contrast requirements with a singular focus against those that address a combination of needs.

Compound Requirement
1. High-priority text shall be presented as sans serif, black alphanumeric characters against a white background, and be at least 18 points (18/72nds of an inch) in height.

Single-Point Requirement
1. High-priority text font shall be sans serif.
2. High-priority text shall be presented as black alphanumeric characters against a white background.
3. High-priority text shall be at least 18 points (Capital "X" height = 18/72nds of an inch).

NO MEANS TO VERIFY

User interface requirements will be difficult, if not impossible, to test when there is no specified or practical way to do so. Sometimes, the root of the problem will be vagueness, as reflected in the following, poorly conceived requirement:

- The software user interface shall be easy to navigate.

The intent of this requirement is clear and sensible; to be sure people can progress through the workflow with ease. But, the requirement does not clearly lend itself to an assessment method or an acceptance criterion. In contrast, the following requirement gets us closer to something that can be tested:

- Each of the workflow screens shall include a title.

Here is another example of an anemic requirement:

- The user interface will protect against use error.

In contrast, the following user interface requirement gets us closer to something that can be verified:

- The user interface shall protect against accidental activation of critical controls.

Further improvement is reflected in the following version:

- Critical controls (i.e., the power switch and emergency stop button) shall be protected from accidental strikes (e.g., inadvertent presses).

NOTES

1 Merriam-Webster.com Dictionary. *pitfall*. www.merriam-webster.com/dictionary/pitfall.
2 Institute for Safe Medication Practices. 2016. *Special Edition: Tall Man Lettering; ISMP Updates Its List of Drug Names With Tall Man Letters*. www.ismp.org/resources/special-edition-tall-man-lettering-ismp-updates-its-list-drug-names-tall-man-letters

5 Writing Top-Quality User Interface Requirements

Here are two polar opposite expressions from different centuries.

PERFECT IS THE ENEMY OF GOOD.[1]

GOOD IS THE ENEMY OF GREAT.[2]

Arguably, there is wisdom in both expressions. The pursuit of perfection should not be a barrier to getting something accomplished. Neither should you settle on good when something better is worthwhile and practical to achieve.

However, when it comes to writing user interface requirements, we advocate for greatness; striving for perfection and perhaps stopping only at the point of greatly diminishing returns.

Our suggestions on how to go from good to great user interface requirements follow.

CREATE AN OUTLINE

Start with a comprehensive outline for the user interface requirements. The outline might organize requirements by component (e.g., hardware, software, labeling) or perhaps by user interface attribute (e.g., text, colors, connections). Regardless of the organization scheme, an outline will help ensure that you do not overlook certain topics.

HOW TO ORGANIZE USER INTERFACE REQUIREMENTS

The requirement organization scheme might or might not be within your control. There might be predetermined categories that are dictated by an existing system. Conforming to a given system, an individual user interface requirement pertaining to the color coding of cables might be grouped with other cabling requirements. Alternatively, a given system might call for all user interface requirements to be placed in a single, consolidated section for integration sake. Both approaches have merit.

The advantage of the first approach (i.e., placing all cabling requirements together) is that a specific user interface requirement may be reviewed in context with related requirements. The advantage of the second approach (i.e., grouping all user interface requirements together) is that it may enable designers to form a more complete, "big picture" of what the product needs

DOI: 10.1201/9781003029717-5

to be from a user interaction standpoint. The ideal system would be one that gives users both views of user interface requirements—distributed and consolidated—by applying content filters. For example, we have seen requirement management systems that are organized by component, but user interface requirements are identified by a "user interface requirement" tag, thereby enabling easy filtering.

If you are going to place user interface requirements in a single section, here are a few, internal organization schemes for your consideration. The key is to organize the content in a way that is most coherent and useful to those who will read the requirements and must meet them.

Option 1—Simplified

- General user interface requirements—addressing design characteristics pertaining to multiple aspects of a product's user interface
- Specific user interface requirements—addressing design characteristics pertaining to a single or limited number of aspects of a product's user interface

Option 2—Component Type-Oriented

- Hardware
- Software
- Packaging
- Labeling

Option 3—Detailed, Adapted, and Expanded from AAMI HE75[3] and Other Sources

- Form[4]
- Displays[5]
- Controls (excluding input mechanisms)[6]
- Digital input mechanisms[7]
- Cables, tubes, and wires
- Connectors and connections
- Transportability
- Cleanability and reprocessing
- Error prevention and recovery
- Labeling[8]
- Software user interfaces
- User assistance (including IFU)
- Alarms[9]
- Sounds (excluding alarms)
- Other sensory elements
- Workstation features
- Signs, symbols, and markings[10]
- Automation

- Interoperability
- Universal design[11]
- Inclusivity[12]
- Branding[13]
- Packaging
- Storage
- Miscellaneous[14]

ADD TECHNICAL ANCHORS

Determine if there is any opportunity to add useful specificity to a requirement without unduly limiting designers' creative freedom when it comes to meeting the requirement. This will also help when the time comes to verify that a design output satisfies a design input (see Chapter 10). For example, while you might require text to be high contrast, it is better to establish a minimum acceptable text-to-background contrast ratio. By stating that "Text shall have a minimum contrast ratio of 7:1,"[15] you have anchored the requirement in a scientifically defensible manner and provided useful guidance to the designer who must implement the requirement.

MAINTAIN A REASONABLE SCOPE

While it is important for a set of user interface requirements to be comprehensive, it might become counterproductive if the requirements become too detailed. For example, it might be enough to specify that on-screen text should be sans serif and ≥12 points to ensure legibility. It might be overkill to specify an alphanumeric character's aspect ratio (height-to-width) and its stroke width (e.g., fine, normal, bold). These details may be left to the good judgment of user interface designers unless there is a particularly strong argument for controlling the ensuing design effort to such a great extent. Such details might also be addressed in a design style guide, rather than as individual user interface requirements.

WRITE CLEARLY AND CONCISELY

Clearly articulated requirements are easier for designers to interpret and implement. As such, user interface requirements should always be reviewed for conciseness and good grammar.

ADD ILLUSTRATIONS

Adding an illustration to a user interface requirement can add clarity without prescribing a particular design solution, which might limit design freedoms. Illustrations may be introduced as an "example only." Some requirement management systems might not enable you to attach illustrations to requirements. In this case, explore alternative ways to capture the illustrations, such as including them in documents that designers will reference when implementing the requirements (e.g., style guide, user interface specification).

ENSURE VERIFIABILITY

As suggested earlier and in Chapter 4, a good user interface requirement is one that facilitates verification. Therefore, ensure that each requirement has a single focus and allows for verification by means of inspection, measurement, or testing in a repeatable manner.

EXAMPLE IMPROVEMENTS

Now, consider the following user interface requirements[16] for an over-the-counter, blood pressure monitor that is intended to be worn on the wrist. Based on the best practices presented earlier, can the requirements be improved upon?

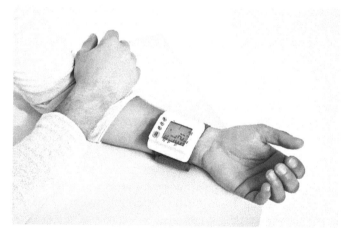

FIGURE 5.1   A blood pressure monitor worn on the wrist.

BEFORE

1. The time of day shall be displayed on the Main Screen.
2. The time of day shall normally be displayed in the 12-hour format.
3. The wrist band shall accommodate a wide range of adults' wrist sizes.
4. The blood pressure reading shall be displayed in the conventional format.
5. The most important information on the main screen should stand out.
6. Buttons should stand out from other user interface features.
7. Units of measure should be indicated.
8. High blood pressure values shall be flagged.

Yes, we believe that each one of these requirements can be improved to some extent, as presented in the following section.

AFTER

1. The time of day (including the hours and minutes as well as the AM or PM indication) shall be displayed on the Main Screen.
2. The monitor shall enable users to switch between a 12- and 24-hour time of day format.
3. The time of day shall default to a 12-hour format including "AM" or "PM," such as shown here.
   - 7:30 AM is early morning
   - 7:30 PM is early evening
4. The monitor shall enable users to change the time of day.
5. The monitor shall require users to confirm the time of day before the change takes effect.
6. The wrist band shall accommodate wrists ranging in circumference from 5.37 inches (first percentile adult female) to 7.74 inches (99th percentile adult male).[17]
7. The blood pressure shall be displayed in the conventional format: with the systolic value positioned above the diastolic value. Example only:

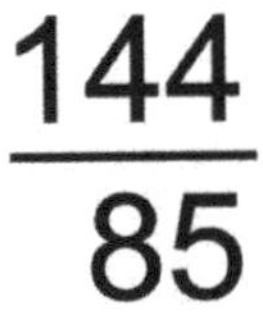

$$\frac{144}{85}$$

FIGURE 5.2 Blood pressure presented in the conventional format (systolic over diastolic).

8. The blood pressure reading shall visually dominate the Main Screen in one or more ways (e.g., relative size, central placement, numeral boldness, spacing away from other content).
9. The blood pressure reading shall be ≥60-point numerals (≥60/72nd inches tall).
10. The blood pressure reading shall be ≥50% larger than any other on-screen, alphanumeric content.
11. Physical buttons shall produce tactile feedback when pressed.
12. On-screen parameters shall have an accompanying text label that includes the units of measure.
13. The monitor shall indicate if a particular blood pressure value is within or outside of the normal, healthy range.

NOTES

1 Attributed to Voltaire (François-Marie Arouet), who reportedly drew the expression from an Italian proverb in his publication called *Dictionnaire Philosophique.* Ratcliffe, S. 2011. *Concise Oxford Dictionary of Quotations.* Oxford: Oxford University Press.

2   Attributed to James Collins. New York City.
    Collins, J. C. 2001. *Good to Great: Why Some Companies Make the Leap . . . and Others Don't.* HarperCollins.
3   Association for the Advancement of Medical Instrumentation. 2009. *ANSI/AAMI HE75–2009: Human Factors Engineering—Design of Medical Devices.* Arlington, VA: Association for the Advancement of Medical Instrumentation.
4   Product size, shape, and style.
5   Displays including physical meters and readouts to virtual representations of the same.
6   Controls including physical buttons and levers to virtual representations of the same.
7   Digital input mechanisms include keypads, touchscreens, computer mice, trackballs, arrow keys, and soft keys.
8   Covering symbols and text appearing on the product itself. Not including information for safety, such as instructions for use.
9   Covers audible, visual, tactile, and any other kinds of alarms.
10 Includes warnings placed on the physical product, displayed on screens, and presented in documents.
11 Product characteristics specifically intended to make products accessible and work effectively for individuals who have different abilities (e.g., individual who has limited hand dexterity).
12 Accommodating various types of people.
13 Pertains to confirming to a corporate branding guide covering such topics as presentation of logo and industrial design.
14 This category may be replaced with one or more additional categories that are tailored to the specific product.
15 Caldwell, B., Cooper, M., Reid, L. G., et al. 2008. *Web Content Accessibility Guidelines (WCAG) 2.0.* W3C. www.w3.org/TR/WCAG20/
16 We have written user interface requirements for a hypothetical blood pressure monitor, which might have a similar appearance to the actual product pictured. The requirements did not come from the product manufacturer.
17 Army Public Health Center. 2020. *Anthropometric Database.* https://phc.amedd.army.mil/topics/workplacehealth/ergo/Pages/Anthropometric-Database.aspx

6 Example User Interface Requirements

ABOUT THE EXAMPLES

This chapter presents 15 sample medical devices along with some possible user interface requirements. The sample devices include home-use products, such as an auto-injector, telehealth app, and cancer screening test kit, as well as devices used in clinical environments, such as a computed tomography (CT) scanner, anesthesia workstation, and hospital bed.

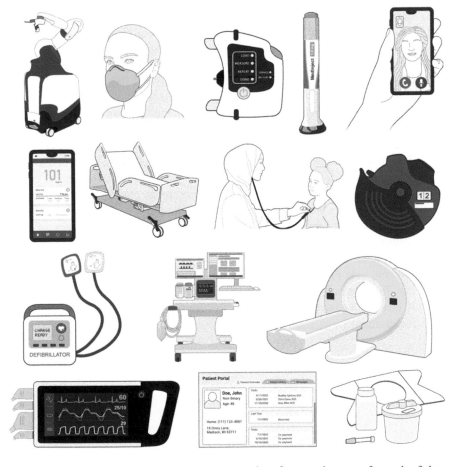

FIGURE 6.1 This chapter presents example user interface requirements for each of these 15 medical products.

DOI: 10.1201/9781003029717-6

Here are a few more things to know about the samples.

- We begin each sample with a product description to give readers a sense for the device's function, users, and use environment.
- The user interface requirements are hypothetical but inspired in many cases by user-friendly design features that we have found or read about in actual products.
- For the sake of keeping the examples clear and concise, we present a dozen or so user interface requirements for each product. We expect that the examples will give readers a sense for the types of user interface requirements that are relevant for a given product, but the examples are far from comprehensive given that actual user interface requirements might number in the hundreds, such as the hypothetical ones we developed for a blood glucose meter (see Chapter 14).
- The examples include many specific, verifiable requirements. That said, we also present some general or high-level requirements that are not sufficiently specific to be verified as is. You can think of each general requirement as a precursor to several more specific requirements that would arise to achieve the goal conveyed by the general requirement. We included a handful of these general requirements to give readers a sense for the design objective driven by a user need, which is not always immediately obvious when reading highly specific requirements. If this was a real use specification, we would discard the general requirements in favor of ones that are verifiable.
- The user interface requirements are intended to be illustrative rather than definitive for the selected, sample medical devices. If we were preparing user interface requirements for an actual product, we would have conducted extensive user research, possibly complemented by other activities described in this book, to arrive at a more definitive set. Therefore, we present the examples for educational purposes only. In this regard, some of the requirements might be off-the-mark from those that might have resulted from the aforementioned research effort.
- We conclude each sample by illustrating a few, possible design solutions that fulfill specific requirements. In many cases, the illustrated solution is just one of many possible ways to satisfy the associated user interface requirement.

STETHOSCOPE

Stethoscopes enable a clinician to listen to patients' heartbeats and breathing. This electronic stethoscope enables recordings and the live transmission of sounds over the Internet.

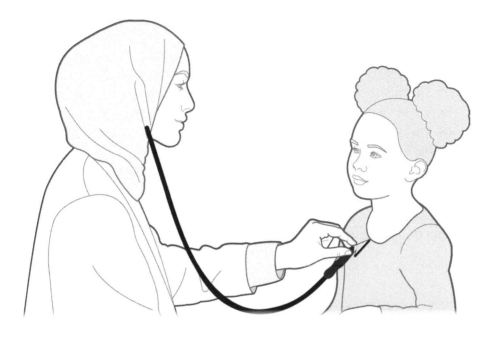

User Interface Requirements

1. The stethoscope's weight shall not exceed 0.5 lb (227 grams).
2. The stethoscope shall require less than 2 seconds to power on and be ready to use.
3. The tubing shall be sufficiently long to drape stably around the back of the neck of a large adult (99th percentile adult male neck circumference = 21.17 inches, 53.78 cm).[1]
4. The tubes shall drape around the user's neck without kinking.
5. The tubes shall be formed from smooth (i.e., nontextured) material.
6. The tubes should be available in multiple colors.
7. The stethoscope shall provide a space for users to write their name on it.
8. The stethoscope shall enable a precision, pinch grip.[2]
9. It shall be possible for users to place the stethoscope on auscultation[3] sites on the patient's body without making hand contact with the patient's skin.

10. The stethoscope shall include a diaphragm cover to serve as an insulating barrier between the metal diaphragm and the patient's skin.

11. The stethoscope shall have an earpiece volume control.

12. The earpiece volume control shall range from 0 dB to 50 dB, enabling the user to set the volume to overcome the ambient noise in hospitals and clinics.[4]

13. The stethoscope shall reduce ambient noise interference from the environment.

14. The stethoscope shall be able to record at least ten, 30-second soundtracks.

15. The stethoscope shall indicate the communication signal strength.

16. The stethoscope shall indicate remaining battery life in four (i.e., 25%) or more increments.

17. The stethoscope shall turn off automatically after 5 minutes of inactivity.

Sample Design Solutions That Fulfill Requirements

Requirement 3
Tube is long enough to wrap around the user's neck and remain in place.

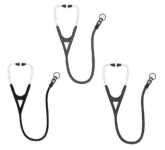

Requirement 6
Tubes are available in black, navy blue, and burgundy.

Requirement 9

The user does not need to contact the patient's skin while holding the device.

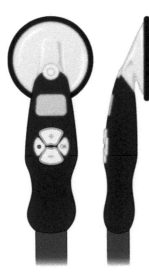

Requirement 11

The stethoscope has "+" and "−" earpiece volume controls.

ANESTHESIA WORKSTATION

This workstation (in various configurations) enables anesthesia providers (e.g., anesthesiologists, certified registered nurse [RN] anesthetists) to put patients under anesthesia, monitor them, and awake them over the course of a medical procedure (typically a surgery). This activity includes delivering anesthetic gases via a breathing circuit and monitoring vital signs presented on one or more displays.

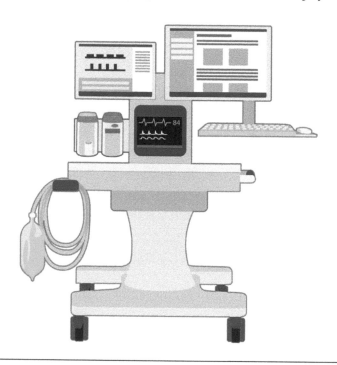

User Interface Requirements

1. The workstation shall have brakes.
2. Brakes shall be applied by means of a single action.
3. The workstation shall weigh less than 200 lb to enable one person to move it from one use location to another.
4. The workstation shall include a work surface suitable for preparing injections and writing notes.
5. The workstation shall include a means to illuminate the work surface.
6. The workstation shall include at least one lockable storage space.
7. The workstation shall provide the means to manage (e.g., gather, organize) cables, wires, and tubes.
8. The brightness of workstation computer displays shall be adjustable.
9. The workstation shall facilitate use on the patient's right and left sides.

10. Anesthetic agent vaporizers shall be within the reach (within 25.2 inches [fifth percentile female]) of the user when the user is seated in a chair directly in front of the workstation.[5]

11. The workstation shall provide the means to attach auxiliary devices (e.g., intravenous infusion pump, syringe pump, optional flowmeter).

12. The workstation shall provide the means to flush the breathing circuit with fresh oxygen with a single action.

13. The oxygen flush control shall be within the user's immediate reach (i.e., ≤14 inches from workstation's front edge and the flush control).

14. The workstation shall redundantly display all critical data.

15. There should be a means to capture a "screenshot" of any display at any time.

16. The workstation should enable institutions to configure the computer-based controls to align with their clinical practices.

Sample Design Solutions That Fulfill Requirements

Requirement 4
The workstation provides a wide working surface enabling the user to perform tasks such as taking clinical notes while also keeping watch on information presented on multiple displays.

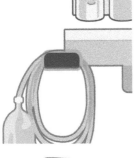

Requirement 7
The workstation includes hooks and brackets to manage cables, wires, and tubes.

Requirement 10
Vaporizers (devices with yellow and blue tops) are placed within the user's reach.

AUTO-INJECTOR

This combination product (a device that contains and delivers a drug) is often used by a layperson to self-administer an injection. Auto-injectors can be used to deliver drugs that treat a wide range of conditions, such as rheumatoid arthritis, Crohn's disease, multiple sclerosis, migraines, and anaphylaxis.

User Interface Requirements

1. The product name shall be printed on the device in ≥12 points.
2. The expiration date shall be printed on the device in ≥12 points.
3. A device containing one dose strength shall be visually distinct (e.g., different color) from devices containing different dose strengths.
4. The needle end shall be visually distinguished (e.g., different shape, color) from other parts of the device.
5. The injection actuation mechanism shall be visually distinguished from the rest of the device.
6. The device's body design shall prevent the device from rolling across flat surfaces.
7. The device shall incorporate a preattached needle rather than requiring the user to attach one prior to performing an injection.
8. The needle shall be hidden from view during all stages of use (before injection, during injection, and after injection).
9. The device shall protect against unintended actuation.
10. The device shall enable users to view the drug contained in the device, thereby enabling visual inspection for particulates and fluid discoloration.
11. The device shall produce a sound when the user commences an injection.
12. The injection duration (i.e., time to deliver the medication, including from needle insertion to needle removal) should be ≤10 seconds.
13. The device shall prevent the user from accessing the used needle after an injection.
14. The device shall prevent reuse.
15. The device shall enable the user to perform the injection step using one hand.

Sample Design Solutions That Fulfill Requirements

Requirements 4, 5

The base (containing the needle) and the injection button have dramatically different appearances. Also, the orange button is evocative of a ballpoint pen's button, suggesting that it provides the means to trigger an injection.

Requirement 6

The device's large, triangular end prevents it from rolling off a flat surface.

Requirement 10

The large medication window enables the user to inspect the medication from multiple angles.

PORTABLE PATIENT MONITOR

This device enables clinicians to monitor a patient at the bedside as well as during transport (e.g., between hospital departments such as the medical/surgical unit and the radiology department). In addition to providing basic readings such as heart rate and oxygen saturation, the monitor also can optionally provide invasive arterial pressure readings.

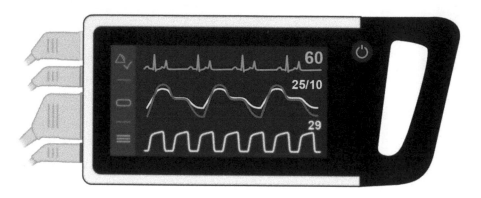

User Interface Requirements

1. Monitor weight shall be ≤3.5 lb (1.59 kg).

2. The monitor shall enable one-handed carrying.

3. The handle shall enable a power grip (fingers wrapped around handle and touching the thumb) by male with a 99th percentile hand length (8.62 inches[6] (21.9 cm)).

4. The monitor shall be able to withstand drops from up to 1 meter (3.3 feet).

5. Connection ports' design shall prevent misconnections.

6. The monitor shall accommodate use of the same cables used by other monitors in the product family.

7. Screens shall produce low levels of emitted light to enable comfortable viewing in a dimly lit room.[7]

8. Touchscreen targets shall cover at least a 0.5-inch (1.27 cm) diameter circle.

9. The touchscreen shall enable operation with gloved hands.

10. The monitor shall enable cleaning using an antiseptic wipe.

11. The battery shall enable up to 5 hours of continuous use.

12. Users shall be able to replace the device's battery while the device is still functioning.

13. Waveforms and their associated numerics shall be color coded.[8]

14. The monitor shall enable institutions to configure information to match their clinical needs (i.e., enable customization).

15. The monitor shall provide the means for institutions to lock the desired information configuration.

16. The monitor shall enable users to connect to a larger monitor to display on-screen information in a larger and more detailed format.

17. The monitor shall connect wirelessly to central stations.

Sample Design Solutions That Fulfill Requirements

Requirement 2
The monitor's small form factor and light weight enables one-handed carrying.

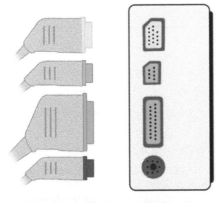

Requirement 5
Color- and shape-coded connection ports and cables limit misconnections.

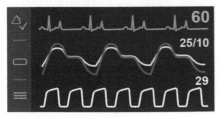

Requirement 7
The dark background limits the light emitted from the monitor, enabling use in a dimly lit room without disturbing the patient.

DIABETES MANAGEMENT APPLICATION

This mobile application (app) works with a glucose meter and desktop computer application to help people with diabetes manage their blood sugar level and communicate results to their healthcare provider.

User interface requirements

1. All primary app functions[9] shall be presented in a main menu.
2. The app shall display text in a sans serif font.
3. The app shall enable users to select a language from a list of those spoken in the target markets.
4. The app shall allow users to track their carbohydrate intake.
5. The app shall remind the user when to perform a blood glucose test according to a preset schedule or testing interval.
6. The app shall remind the user when to take insulin according to a preset schedule or testing interval.
7. The app's resting screen shall display the most recent, date-stamped glucose measurement.
8. The app shall enable users to annotate glucose readings with preset notes about associated insulin intake, food intake, and exercise.
9. The app shall allow pairing with ≥2 meters that may be in use by a single user.
10. The app shall maintain a log of glucose readings taken by paired glucose meters.
11. The app shall give users a "big picture" view of blood glucose readings that may be within and/or outside of range.

12. The app shall indicate blood glucose level patterns (e.g., periods of being out of range) that might indicate the need for the user to adjust her/his/their routine.

13. The app shall enable users to view blood glucose readings within a selected day or week.

14. The app shall provide users with average blood glucose levels for the past two weeks, four weeks (one month), three months, six months, and one year.

15. The app shall enable users to share their data securely with trusted individuals via an email or text message.

Sample Design Solutions That Fulfill Requirements

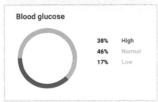

Requirement 7

The main screen displays the most recent blood glucose measurement at the center or the top menu.

Requirement 11

These two graphics reveal a pattern of some blood glucose readings being below, at, and above target range values.

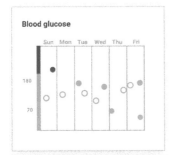

Requirements 12 and 13

This screen enables users to see the range of blood glucose readings from the past 8 days, highlighting the lows (blue numbers) and highs (pink).

CT SCANNER

This device enables clinicians to see inside the human body by first taking X-rays from various angles and then generating cross-sectional images showing anatomical features (e.g., bones, organs, blood vessels, soft tissue) in considerable detail.

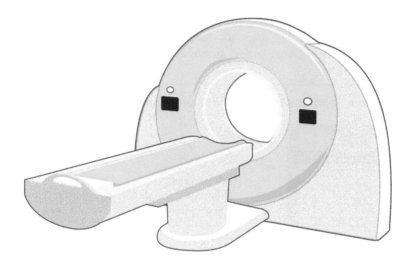

User Interface Requirements

1. The patient table shall have removable padding.
2. The patient table padding shall be impervious to fluids.
3. The patient table shall have attachment points for patient restraints placed at multiple locations such as near the patient's shoulders, waist, and knees.
4. The patient table shall incorporate storage for restraints and other accessories (e.g., body part positioning and immobilization devices).
5. Patient table motion controls shall be accessible (i.e., ≤16 inches) from both sides of the patient table.
6. The scanner shall have emergency stop control(s).
7. The emergency stop control(s) shall be accessible (i.e., ≤16 inches) from both sides of the patient table.
8. The scanner shall enable continuous, duplex communication between the operator (located in the control room) and the patient.
9. The scanner shall enable patients to listen to audio (e.g., music, intercom messages from the healthcare professionals operating the scanner).
10. The bore shall accommodate large (including obese) individuals; minimally an adult male with a 99th percentile waist circumference (41 inches).

11. The scanner shall safeguard the patient from physical trauma due to scanner component motion.

12. The scanner shall safeguard the operator from physical trauma due to scanner component motion.

13. The bore surface shall enable items (e.g., clothing, wires, tubes) to move through it without snagging.

Sample Design Solutions That Fulfill Requirements

Requirement 1
Patient table padding (in green) is removable, thereby enabling easy cleaning.

Requirement 6
An emergency stop control is above the touch screens that are positioned on each side of the patient table.

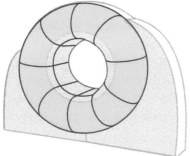

Requirement 13
The bore has a smooth surface and tapered form to protect the patient against lacerations and abrasions as well as to avoid snagging moving elements.

CANCER DETECTION TEST KIT

This home-use kit includes the materials needed to collect a stool specimen, preserve the specimen, and mail it to a lab for colorectal cancer screening.

User Interface Requirements

1. The kit's expiration date month shall be presented using letters (e.g., "AUG").
2. The kit's expiration date shall be printed on each kit component (i.e., sample container, tube, bottle, and bracket).
3. The kit shall enable the person being tested to perform the collection procedure independently (i.e., it shall be a one-person task).
4. The entire sampling procedure shall not require materials, accessories, or tools other than those provided in the kit, except for standard home fixtures (i.e., toilet, sink, and soap).
5. The IFU shall be written in both English and Spanish (North American market versions).
6. The IFU shall graphically illustrate the steps.
7. The IFU shall refer users to an online video that also provides instructions.
8. The return shipping container shall be preaddressed.
9. Before instructing users to collect a sample, the IFU shall advise users to collect a sample at a time when they can ship it back to the lab within a day of collection.
10. The IFU shall state why it is important to ship the sample within one day of collection.
11. The sample collection container shall have a diameter of ≥8 inches to ensure effective stool sample collection with a minimal chance of spillover.

12. The sample collection container shall enable users to position it securely within the bowls of toilets ranging from 12 to 16 inches wide.

13. The stool sample probe shall enable the user to scrape[10] the stool sample without contaminating her/his hand.

14. The stool sample probe shall provide visual cues that it has collected a sufficient sample.

15. Kit container caps/lids shall incorporate elements that facilitate opening and closing.

16. The kit's outer appearance shall not indicate its purpose to ensure discreetness.

Sample Design Solutions That Fulfill Requirements

Requirement 7
A 4-minute video, made available on the Internet, demonstrates how to use the kit.

Requirement 9
Instructions prompt users to consider whether they can return the sample within one day of collection *before* instructing users on how to collect the sample. Presenting the information in this order decreases the likelihood that users collect samples at a time when they are unable to promptly return the sample.

Requirement 12
Bracket, which is designed to straddle the sides of toilet bowls of various width, suspends the sample container at the bowl's center.

Requirement 15
Sample container lid has raised bars that facilitate twisting it off of and onto the container securely.

DRY POWDER INHALER

This device enables people to inhale a dry powder medication to treat chronic obstructive pulmonary disease.

User Interface Requirements

1. The inhaler shall prevent mouthpiece contamination when the inhaler is not in use.

2. The mouthpiece cover shall remain connected to the inhaler to ensure users do not lose the cover.

3. The mouthpiece cover shall enable users to perform an inhalation without their mouth touching the cover (i.e., the cover shall not obstruct the user from performing an inhalation).

4. The mouthpiece's design (e.g., texture) shall enable the user to seal[11] his/her/their lips on the mouthpiece.

5. The inhaler shall display the number of doses remaining.

6. The number of remaining doses shall be presented as black numerals on a white background or a similar, high-contrast pairing.

7. The number of doses remaining shall decrement immediately after the device delivers a dose.

8. The inhaler shall provide audible feedback when the dose counter decrements.

9. The dose counter shall be visible when the mouthpiece cover is open.

10. The dose counter shall be visible when the mouthpiece cover is closed.

11. The inhaler shall require cleaning ≤1/month during the expected period of use.

12. The mouthpiece shall enable the user to clean it using a wipe.

13. The inhaler shall incorporate a legible brand name label that is at least 4.3mm tall.[12]

14. The inhaler shall enable the user to record the date of first use on the inhaler.

15. The device shall provide feedback (e.g., tactile, auditory) when the user positions the cover over the mouthpiece.

Sample Design Solutions That Fulfill Requirements

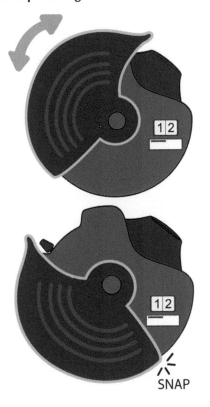

Requirement 3
The mouthpiece cover pivots and stays out of the way because it snaps into place.

Requirements 4
The mouthpiece has a smooth, rounded shape, which supports a good seal with the user's lips.

Requirement 5

The inhaler displays the number of doses remaining.

ELECTRONIC MEDICAL RECORD

This product is a multifaceted, cloud-based, and customizable software application that enables healthcare institutions and healthcare providers to input, store, and retrieve health information about their patients as well as perform ancillary tasks, such as scheduling patient visits.

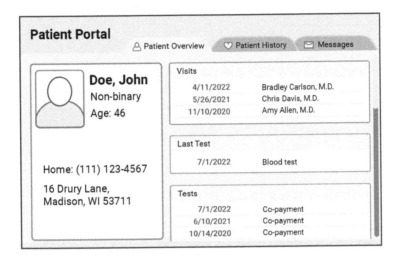

User Interface Requirements

1. All screens containing patient data shall prominently (i.e., ≥16-point text) display the patient's name.

2. The primary patient data screen shall enable the inclusion of a patient photograph.

3. Each screen's title shall indicate the screen's primary function.

4. Major, selectable options[13] shall be displayed as an icon reinforced by a text label.

5. Text in lists and paragraphs shall have a consistent alignment (e.g., all left-justified).

6. Data field labels shall have a different appearance (e.g., different color) than the data itself.

7. Unavailable (i.e., inactive) options shall have a different appearance than available options.

8. Text and backgrounds shall have a contrast ratio of ≥7:1.

9. Screens shall use ≤7 colors.

10. Important and safety critical data inputs and actions shall require user confirmation.

11. Screen response to input lag time shall be ≤40 milliseconds.
12. Screen prompts shall be placed in a consistent location.
13. Buttons shall be spaced such that their centers are separated by 0.75 inches.[14]
14. Text size shall be adjustable.

Sample Design Solutions That Fulfill Requirements

Requirement 2
Space is allotted for a patient photograph to support correct identification.

Requirement 4
Inclusion of an icon and associated text label "Patient History" makes the major feature visually distinct from other screen content.

Requirement 10
Dialog box requires the user to save changes or cancel before continuing with other tasks.

ROBOTIC ARM AND SOFTWARE

This device enables orthopedic surgeons to make precise bone cuts in the course of performing partial knee, total knee, and total hip procedures, thereby improving the positioning of implants and the overall outcome of these orthopedic procedures.

User Interface Requirements

1. The system shall be portable (i.e., weigh ≤200 lb), enabling users to move it among several use, maintenance, and storage locations.

2. The system shall have brakes, or an equivalent means, to lock it in place, thereby preventing unwanted movement.

3. All brakes shall be applied and released by means of a single control action.

4. The system shall provide the means for the user to grip the robotic arm to stabilize it.

5. The system shall enable component implantation within 2° of the target position.

6. The system shall enable the surgeon to make intraoperative adjustments (e.g., anterior/posterior component translation, flexion/extension adjustments) to bone cutting actions.

7. The system shall enable users to virtually plan the procedure, including virtually placing an implant within the patient, before executing the procedure.

8. The system shall indicate cutting boundaries without the use of physical cutting guides.

9. The system shall provide multichannel, sensory feedback when a tool has been securely attached.

10. Using the system shall not increase the overall time required to perform robot-assisted surgical procedures as compared to unassisted procedure times.

11. The system shall provide the means to immediately lock (i.e., freeze) the position of all moving components.

12. In the event that the user locked the position of all moving components, the system shall subsequently enable the user to withdraw tools manually (e.g., robotic arm with attached saw blade) from the surgical site.

13. The drape shall cover the robotic arm from saw to joint, at minimum.

14. The drape shall enable users to access grip points on the robotic arm.

Sample Design Solutions That Fulfill Requirements

Requirement 1

Four casters enable users to adjust the console's position as well as transport it among use, maintenance, and storage locations.

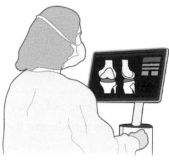

Requirement 7

System software enables surgeons to plan the surgery and determine appropriate anatomical reference points, which are essential to operating precision prior to preforming the procedure.

Requirement 14

The custom drape kit covers the hand hold while still enabling the surgeon to wrap his or her fingers securely around the hand hold.

AUTOMATIC EXTERNAL DEFIBRILLATOR

This device instructs laypersons on how to deliver effective cardiopulmonary resuscitation (CPR) and how to administer a cardioverting shock to someone experiencing a cardiac event.

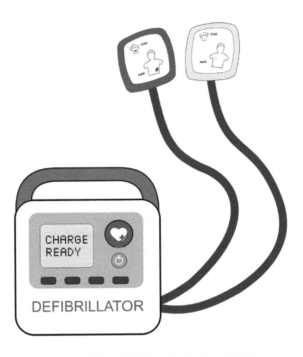

User Interface Requirements

1. The AED shall have a carrying handle.
2. The AED shall provide voice guidance for attaching the electrodes.
3. The AED shall provide voice guidance for delivering a shock.
4. Voice guidance shall use simple words that can be understood by individuals with an eighth-grade or higher reading level.
5. Voice guidance shall direct the user to choose carefully between a set of adult versus pediatric electrode pads.
6. Adult and pediatric electrode pads shall be visually distinct from one another.
7. Electrode pads shall be disposable.
8. Each electrode pad shall graphically indicate its proper placement on the patient.
9. The AED shall provide on product, graphical IFU.
10. Graphical instructions shall be legible in dim lighting conditions (as low as 50 lux).[15]

11. The AED shall have a large (≥72 points) indication of its purpose.

12. Users shall be instructed both verbally and graphically not to touch the patient when a shock is pending or occurring.

13. The AED shall have no detachable parts except for the disposable electrode pads.

14. The AED will provide the means for the user to confirm that it is fully functional (i.e., ready for use, including having a proper electrical charge).

15. The AED shall direct users to call 911 to summon rescue services.

Sample Design Solutions That Fulfill Requirements

Requirement 8
Graphics indicate where each pediatric electrode should be placed on small versus larger children.

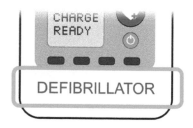

Requirement 11
The front of the device clearly indicates its purpose: DEFIBRILLATOR.

Requirement 15
An LCD display states "CHARGE READY," suggesting it is functioning properly and has enough electrical charge to treat a patient.

TONOMETER

This tonometer enables people at risk of high intraocular pressure (IOP), such as those with or developing glaucoma, to self-test their IOP on a daily basis to judge the need for, and the effectiveness of, medical treatment. Patients hold the device up to their eye, and a disposable probe gently and swiftly tap on the eye's surface to measure the IOP.

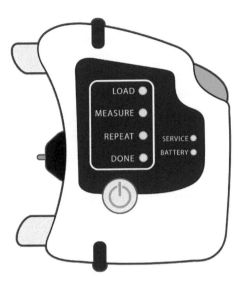

User Interface Requirements

1. The device shall enable one-handed use.

2. The device shall be equally suited for right- and left-handed use.

3. The device's outer material shall have a nonslip texture (e.g., material with a high coefficient of friction)[16] to help users maintain a secure grip.

4. The device shall accommodate human faces of different shapes and sizes.

5. Probes shall be protected from contamination until they are attached to the tonometer.

6. The probe insertion port shall guide the probe into place, thereby not requiring fine motor control by the user.

7. The device shall prevent premature, unintended contact with the eye, ensuring contact only by the moving probe at the right point in time.

8. The device shall enable users to start an IOP measurement without having to change their grip on the device once it is properly positioned in front of the eye.

9. The device shall guide users to gaze in the correct direction (i.e., directly toward the probe), thereby enabling a successful measurement at the cornea's center.

10. The device shall enable users to hold it with a neutral wrist position[17] during the course of performing a measurement.

11. The treatment shall require users to hold the device up to their face for ≤10 seconds.

12. The device shall provide positive feedback when it takes a successful measurement.

13. The device shall provide feedback when it fails to capture a measurement and the measurement must be repeated.

14. The device shall protect against the reuse of a used probe.

15. The device shall come with a storage case.

Sample Design Solutions That Fulfill Requirements

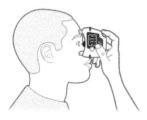

Requirement 7
Adjustable, standoff features prevent device contact with the eye, allowing only the probe to make contact during a measurement.

Requirement 8
The measure button is easily accessed by the index (pointer) finger.

Requirement 9
The probe is centered within a green circle indicating exactly where users should look during the measurement.

Requirement 15

The portable storage case contains tonometer, probes, and instructions for use.

HOSPITAL BED

This bed is designed for use in intensive care environments. It is designed to accommodate a wide range of patient body positions, prevent or treat certain pulmonary conditions linked to patient immobility, and support early patient mobility.

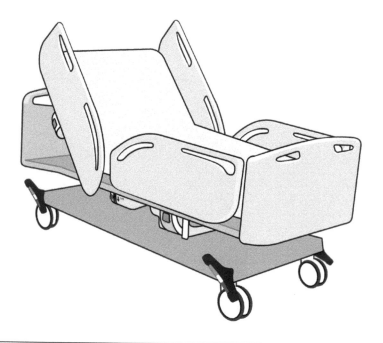

User Interface Requirements

1. Siderails (guardrails) shall prevent body part entrapment.
2. Siderail (guardrail) motion shall be damped to move at a slow pace instead of "free falling" when released from a raised position.
3. It shall be possible to reposition siderails (guardrails) using only one hand.
4. Brakes shall be accessible from both sides, the front, and the foot of the bed.
5. All brakes shall be locked or unlocked via a single control action.
6. Brakes shall provide a visual indication of whether they are locked versus unlocked.
7. Bed steering controls shall enable the bed to be moved with either two or four wheels able to freely rotate, thereby facilitating relatively straightforward motion or lateral translation, respectively.
8. Both ends of the bed shall provide a means to manually push and pull the bed.
9. It shall require a single action to place the bed in a configuration suitable for the delivery of CPR.

10. Displays should automatically dim or turn off after use to prevent disturbing patients by illuminating a normally dim or dark room with light.

11. Bed shall indicate its upper body/head angle.

12. Bed movement controls shall require continuous input to produce movement.

13. There shall be a means to lock controls so that the bed cannot be repositioned by the patient or visitors.

Sample Design Solutions That Fulfill Requirements

Requirement 1
Siderail geometry precludes large gaps that could entrap a patient's head, arm, or leg.

Requirement 3
Siderail release mechanism enables one-handed control of the component's position (raised versus lowered).

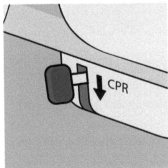

Requirement 9
CPR control can be actuated quickly, with a single action, using a hand or foot.

TELEHEALTH APP

This product is a mobile medical application (app) used to conduct telehealth visits. The patient enters his or her demographic details and a description of the reason for the visit (e.g., symptoms, concerns) into the app. Then, the application connects the patient to a video chat with a physician or psychologist. The app also enables the physician/psychologist to record visit notes and prescribed treatments.

User Interface Requirements

1. The app shall require the patient to confirm his/her/their demographic details (e.g., name, date of birth, phone number) when scheduling a visit.

2. Each screen shall have a meaningful title that indicates its primary function.

3. Selection lists (e.g., list from which the patient selects their symptoms) longer than ten items shall be divided into subsections (i.e., chunks) with descriptive subsection titles.

4. Subsection titles within lists shall persist at a subsection's top until the user scrolls to the next subsection.

5. The app shall visually distinguish selection fields that allow a single selection from selection fields that allow multiple selections.

6. Within selection lists, selected items shall be differentiated from unselected items.

7. Data entry fields shall initially be presented as empty rather than filled-in with default values.

8. Times shall be presented in the 12-hour format (most commonly used in the United States, for example).

9. "AM" and "PM" labels shall be displayed in all-capital letters.

10. Dates shall be presented using the format MM/DD/YYYY (i.e., month followed by day followed by year) with single digit entries preceded by a zero (e.g., September = 09)

11. If a screen transition will take more than 1 second, the screen shall present a dynamic, visual indication that it is in active transition (i.e., loading).

12. The app shall provide visit status updates to the patient (i.e., estimated wait time, physician reviewing chart, and physician ready to start visit).

13. The estimated wait time shall be presented before the patient is prompted to confirm a visit.

14. The app shall prompt the patient to confirm that she/he/they are ready to communicate with a healthcare provider before starting the telehealth visit (i.e., establishing a connection).

15. The past visit list shall include the day, month, and year of each past visit.

Sample Design Solutions That Fulfill Requirements

Requirement 3
Selection list includes subsection titles, enabling the patient to easily scan the list and locate relevant symptoms.

Requirement 4
Subsection titles within lists shall persist at a subsection's top until the user scrolls to the next subsection.

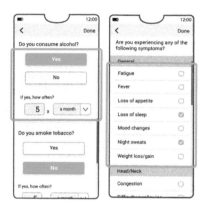

Requirement 5

Check boxes denote fields that allow for multiple responses, whereas single response fields have a different appearance.

RESPIRATOR

This N95 respirator is a face covering that became familiar to many during the COVID-19 pandemic. It serves to protect against the transmission of airborne contaminants, such as viruses, from one person to another. It complies with filter performance guidelines set by the U.S. National Institute for Occupational Safety and Health (NIOSH).

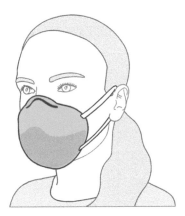

User Interface Requirements

1. The respirator shall indicate on its exterior that it is a NIOSH-approved N95 respirator.
2. The respirator shall be available in ≥3 sizes to enable users with faces of different shapes and sizes to achieve a tight seal.
3. The respirator shall enable users to adjust the fit around the nose.
4. The means to hold the respirator in place (e.g., ear loops) shall enable extended uses (donning and doffing 100 times) without a reduction in their functional capability.
5. The respirator's pressure against the face shall be relatively evenly distributed around the respirator's perimeter (i.e., not concentrated in a couple of places).
6. Breathability (differential pressure) shall be <2.1, which helps ensure the perception of the mask being "cool."
7. The respirator shall resist moving out of its optimal position (i.e., riding up or down on the face more than 1 cm).
8. Material touching the face shall have what users regard to be a comfortable feel (i.e., smooth, soft).
9. An unused respirator shall have no discernable odor.
10. The respirator shall readily reveal blood contamination (i.e., its exterior color shall have a contrast ratio of 3:1 with dark red [hex #8b0000]).

11. The respirator shall be provided with IFU.

12. The IFU shall indicate the proper means to don and doff the product safely.

13. The IFU shall indicate when to discard the respirator.

14. The IFU shall be available online.

15. The IFU shall be available in a mobile version (i.e., viewable on a smartphone).

Sample Design Solutions That Fulfill Requirements

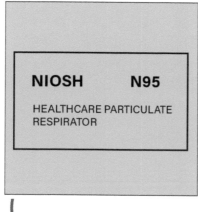

Requirement 1

The respirator has NIOSH N95 printed on the visible (out facing) side.

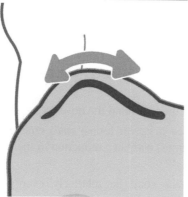

Requirement 3

The aluminum nose wire enables the user to adjust the fit around the bridge of the nose.

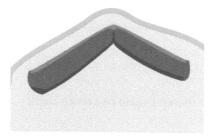

Requirement 8

Foam padding creates a soft seal around the bridge of the nose.

NOTES

1 Fullenkamp, A. M., Robinette, K. M., and Daanen, H. A. 2008. *Gender Differences in NATO Anthropometry and the Implication for Protective Equipment.* Air Force Research Lab Wright-patterson AFB OH Biomechanics Branch.

2 The precision pinch grip is where an object is pinched between the palmar surface of the fingers and the opposing thumb.

3 Auscultation is the term for listening to the internal sounds of the body.

4 Association for the Advancement of Medical Instrumentation. 2009. *ANSI/AAMI HE75–2009: Human Factors Engineering—Design of Medical Devices.* Arlington, VA: Association for the Advancement of Medical Instrumentation.

5 Attwood, D. A., Deeb, J. M., & Danz-Reece, M. E. 2004. *Ergonomic Solutions for the Process Industries.* Amsterdam: Elsevier.

6 Gordon, Claire, Thomas, C., Clauser, C., & Bradtmiller, B. 1988. *Anthropometric Survey of U.S. Personnel: Summary Statistics Interim Report for 1988.* https://apps.dtic.mil/sti/citations/ADA209600

7 Dark mode design tips: https://developer.apple.com/design/human-interface-guidelines/ios/visual-design/dark-mode/

8 Matching waveform and numeric color is conventional for patient monitors.

9 Primary app functions are defined in the design specification.

10 To use the kit, users must scrape the stool sample using a probe to transfer a small amount of the sample into a test tube.

11 Seal is defined as a mouthpiece design that has created no gaps between mouthpiece and lips.

12 Association for the Advancement of Medical Instrumentation. 2009. *ANSI/AAMI HE75–2009: Human Factors Engineering—Design of Medical Devices.* Arlington, VA: Association for the Advancement of Medical Instrumentation.

13 Features defined as "major options" are listed in the design specification.

14 Association for the Advancement of Medical Instrumentation. 2009. *ANSI/AAMI HE75–2009: Human Factors Engineering—Design of Medical Devices.* Arlington, VA: Association for the Advancement of Medical Instrumentation.

15 Health and Safety Executive. n.d. *Human Factors: Lighting, Thermal Comfort, Working Space, Noise and Vibration.* www.hse.gov.uk/humanfactors/topics/lighting.htm

16 Association for the Advancement of Medical Instrumentation. 2009. *ANSI/AAMI HE75–2009: Human Factors Engineering—Design of Medical Devices.* Arlington, VA: Association for the Advancement of Medical Instrumentation.

17 A neutral wrist position is where the wrist is in straight alignment with the forearm (i.e., no flexion, extension, and radial or ulnar deviation). The wrist is at the mid-point between supination and pronation. This position is commonly called the handshake position.

7 Conducting Research to Inform User Interface Requirements

How big an effort could it be to write a comprehensive set of user interface requirements? Well, there are two main variables. One is the breadth of the research you are prepared to conduct in preparation to write the user interface requirements. Another is the characteristics of the device under development.

One way to think about the breadth of research with regard to writing user interface requirements is to think of your school days—be they long ago or still in progress—when you were tasked with writing a report, term paper, thesis, or dissertation. You could probably produce an outline for the content before performing a significant amount of research. Also, you might be able to "block in" some basic content based on your existing knowledge. However, it is unlikely you would be able to produce the entire document—or at least a good one filled with new insights—without conducting at least a modicum of research. The same principle applies to writing user interface requirements.

Therefore, when you scope the user interface requirement writing effort, you need to consider the amount of the research that you will need to do to support the effort.

FIGURE 7.1   The level of effort to develop user interface requirements can be low or high, depending on the breadth of the research effort and the characteristics of the device.

DOI: 10.1201/9781003029717-7

The FDA's HF guidance and IEC 62366 provide a good running start at knowing what general types of research to perform and then leave it to the manufacturer to decide about the specifics.

LEARN ABOUT USERS

The commonly used methods to learn about users' needs and preferences are the following.

OBSERVATIONS

Observations of people engaged in relevant use scenarios and performing relevant tasks. There are multiple ways to approach this research that vary in terms of formality, data collection technique, and scale. Methods include unobtrusive observation of people engaged in activities of interest (i.e., ethnographic research) and interviewing people engaged in the same type of activities (i.e., contextual inquiry).

DEFINITIONS: TASKS VERSUS USE SCENARIOS

Task = Action performed by a user to achieve a specific goal
Use scenario = Group of tasks organized in a logical order to represent a natural workflow

Example

Summary of observation activity: Observe healthcare professionals perform the use scenario: "Prepare for and administer an injection to a patient"

Device: Prefilled Syringe

Tasklist comprising complete use scenario:

1. Gather injection supplies onto a clean, flat surface
2. Inspect prefilled syringe for cracks/damage, an appropriate expiration date, and medication clarity
3. Wash hands
4. Select an injection site (upper thighs or abdomen)
5. Disinfect injection site with an alcohol wipe
6. Remove the needle cap
7. Eject air by pushing the plunger until a drop of medication appears
8. Squeeze the skin around the injection site and hold
9. Insert the needle into the squeezed skin at a 45° angle
10. Press the plunger slowly until all medication has been delivered
11. Remove the needle from the skin
12. Discard the prefilled syringe and needle cap in a sharps container

INDIVIDUAL INTERVIEWS

Individual interviews may be considered the gold standard for how to learn about users. The key is to talk to the right people and enough of them to develop an accurate sense of their characteristics, needs, and preferences. Another key is to conduct an unbiased interview; one that does not consciously or unconsciously shape the participants' answers. For example, someone working on a mobile ultrasound workstation might objectively ask, "Have you ever experienced physical ailments related to repetitive body movements?" It would be less objective to ask, "Do you think the keyboard position should be adjustable so that it doesn't give you Carpal Tunnel Syndrome?" That said, after getting the answer to the first, unbiased question, the researcher could ask the follow-up question, "Do you have any recommendations regarding keyboard design and placement?" If the answer was, "I'd like to see it at a comfortable height," the researcher could simply say, "Tell me more." Ultimately, after a discussion that is deliberately nonleading, the researcher could then ask the direct question, "What, if any, keyboard position adjustments would be desirable and why?"

GROUP INTERVIEWS

Group interviews can serve the same purpose as individual interviews with the added benefit of speeding up the process because you can talk to eight people at once for a couple of hours as opposed to talking to eight people individually for an hour a piece. You just have to be sure that all interviewees get a chance to express themselves and that the group dynamic (e.g., one individual with strong opinions could influence others' willingness to express differing opinions) does not skew the findings. Often, group interview participants can draw a consensus regarding desirable product features, replacing the need for researchers to assimilate the findings from individual interviews to draw a potentially less reliable conclusion about how to prioritize features.

LITERATURE SEARCHES

Literature searches can uncover a bounty of information about users. For example, you can learn much about users by reading occupational descriptions provided on government and industry association websites. For instance, the following description of the RN role tells you a lot that can be translated into user interface requirements.

Excerpt from *U.S. Bureau of Labor and Statistics' Occupational Outlook Handbook* (www.bls.gov/ooh/healthcare/registered-nurses.htm):

Summary:

RNs provide and coordinate patient care and educate patients and the public about various health conditions. RNs usually take one of three education paths: a bachelor's degree in nursing, an associate's degree in nursing, or a diploma from an approved nursing program. RNs must be licensed.

Duties:

RNs typically do the following:

- Assess patients' conditions
- Record patients' medical histories and symptoms
- Observe patients and record the observations
- Administer patients' medicines and treatments
- Set up plans for patients' care or contribute information to existing plans
- Consult and collaborate with doctors and other healthcare professionals
- Operate and monitor medical equipment
- Help perform diagnostic tests and analyze the results
- Teach patients and their families how to manage illnesses or injuries
- Explain what to do at home after treatment

Most RNs work as part of a team with physicians and other healthcare specialists. Some RNs oversee licensed practical nurses, nursing assistants, and home health aides.

RNs' duties and titles often depend on where they work and the patients they work with. For example, an oncology nurse works with cancer patients and a geriatric nurse works with elderly patients. Some RNs combine one or more areas of practice. For example, a pediatric oncology nurse works with children and teens who have cancer.

Injuries and illnesses:

RNs may spend a lot of time walking, bending, stretching, and standing. They are vulnerable to back injuries because they often must lift and move patients.

The work of RNs may put them in close contact with people who have infectious diseases, and they frequently come into contact with potentially harmful and hazardous drugs and other substances. Therefore, RNs must follow strict guidelines to guard against diseases and other dangers, such as accidental needle sticks and exposure to radiation or to chemicals used in creating a sterile environment.

Work schedules:

Nurses who work in hospitals and nursing care facilities usually work in shifts to provide round-the-clock coverage. They may work nights, weekends, and holidays. They may be on call, which means that they are on duty and must be available to work on short notice.

Nurses who work in offices, schools, and other places that do not provide 24-hour care are more likely to work in regular business hours.

LEARN ABOUT USE ENVIRONMENTS

The methods described earlier are also useful for expanding your understanding of the environments in which a product is used. Common methods to learn about use environments that can be translated into user interface requirements include the following.

Site Visits

Site visits are a straightforward way to gain insights leading to user interface requirements. During a visit, you might be able to take photos, note the types of equipment in use, measure spaces, and measure lighting and sound levels, among other activities, all to characterize the use environment. Of course, this will require the facility's permission and privacy protection. Among many other discoveries, researchers working for a mobile ventilator manufacturer (for example) might learn that there is very little room for a hospital bed, mobile ventilator, and two attendants in some hospital elevators. This discovery might lead them to limit the device's depth without making it unstable.

Individual Interviews

Individual interviews focused on user needs can be extended to include a discussion about use environment characteristics. Such discussions can generate insights that might not be self-evident during a site visit. For example, a nurse who works in an intensive care unit might describe situations in which small patient care areas are congested with equipment, and it becomes difficult to reach behind a device for a power switch to turn it on.

Group Interviews

Group interviews focused on user needs can also be extended to include a discussion about use environment characteristics and offer the aforementioned advantage of generating consensus regarding the features of a product that will work well in the particular environment.

Literature Reviews

Literature reviews can also be a plentiful source of information about use environments. This is illustrated in the following excerpt of an article about a hospital's central laboratory where lab technicians perform blood tests.

> Excerpt from Wikipedia's description of a Medical Laboratory, as of 2021:[1]
> A medical laboratory or clinical laboratory is a laboratory where tests are carried out on clinical specimens to obtain information about the health of a patient to aid in diagnosis, treatment, and prevention of disease. Medical laboratories vary in size and complexity and so offer a variety of testing services.
> In a hospital setting, sample processing will usually start with a set of samples arriving with a test request, either on a form or electronically via the laboratory information system (LIS). Specimens are prepared for analysis in various ways. For example, chemistry samples are usually centrifuged and the serum or plasma is separated and tested. Many specimens end up in one or more sophisticated automated analyzers, that process a fraction of the sample to return one or more test results. Some laboratories use robotic sample handlers (Laboratory automation) to optimize the workflow and reduce the risk of contamination from sample handling by the staff.

The work flow in a hospital laboratory is usually heaviest from 2:00 am to 10:00 am. Nurses and doctors generally have their patients tested at least once a day with common tests such as complete blood counts and chemistry profiles. These orders are typically drawn during a morning run by phlebotomists for results to be available in the patient's charts for the attending physicians to consult during their morning rounds. Another busy time for the lab is after 3:00 pm when private practice physician offices are closing. Couriers will pick up specimens that have been drawn throughout the day and deliver them to the lab. Also, couriers will stop at outpatient drawing centers and pick up specimens. These specimens will be processed in the evening and overnight to ensure results will be available the following day.

CONVERTING USER INPUTS INTO REQUIREMENTS

Let's examine a set of user research findings and how they can be converted into user interface requirements. In this case, assume that group interviews with people who test their blood sugar levels using a glucose meter have stated their preferences regarding a next generation meter.

FIGURES 7.2–7.4   Three sample glucose meters that use test strips.

Here are five quotations from the hypothetical interviews:

The readout should be as big as possible, ideally so I could read it without putting on my reading glasses.

It gets confusing if there are too many buttons. I'd rather the device have a few large buttons that I can tell apart from each other and that are easy to press.

I'd love it if you didn't need test strips because I'm always dropping a strip, putting it in the meter the wrong way, and applying too much or too little blood to it.

I assume it will have a display. So, if it does, it would be nice if was bright, even in the daytime, so that the information is easy to read.

It would be nice if the meter could send results to my iPhone without me ever having to do anything. That would save me time and I wouldn't have to remember to do it.

We have written these to sound the way people speak, which is not always grammatically perfect. But, direct quotations give you a nice feel for what the respondent had in mind.

User inputs can be quite specific, such as to make the readout large. Others describe a design characteristic but are not so prescriptive as to suggest a display backlight, for example. In almost all cases, such inputs cannot stand as user interface design requirements. Rather, the ones that represent a consensus among users and align with professional judgments must be converted into a useful form.

Created by Anatolii Babii
from Noun Project

FIGURE 7.5   Group interview.

Created by Justin Blake
from Noun Project

FIGURE 7.6   Write requirements.

Here are some user interface requirements derived from the earlier quotations. Notice that a single user input might have several parts to it, leading to multiple, discrete user interface requirements.

1. Glucose values shall be displayed in 60 points or larger font.
2. The mechanical buttons' top shall be concave to "capture" the fingertip.
3. At a minimum, the mechanical buttons' top shall cover a 0.5-inch diameter circle.
4. Mechanical buttons shall have at least 2.5 mm of travel to provide tactile feedback.
5. Test strips shall indicate which end is inserted into the meter.
6. Test strips shall indicate where blood should be applied to it.
7. It shall be possible to insert a test strip into the meter with either surface of the strip facing up.
8. Test strips will work properly (i.e., produce a test result, not an error) even if the user applies an excessive amount (5 µL) of blood to it.
9. The display shall be backlighted.
10. The backlight shall turn on whenever the meter is in use.
11. The meter shall automatically upload test results to a paired smartphone.

Notice that the user interface requirements do not exactly match what users suggested. For example, it might not be possible or commercially viable to produce a meter that does not require test strips. But, the respondents' interest in eliminating test strips can be addressed by requirements that make the strips easy to use and less likely to induce use errors.

NOTE

1 Source: https://en.wikipedia.org/wiki/Medical_laboratory#:~:text=A%20medical%20laboratory%20 or%20clinical,treatment%2C%20and%20prevention%20of%20disease.&text=Medical%20 laboratories%20vary%20in%20size,a%20variety%20of%20testing%20services

8 Identifying Risks to be Mitigated through Design

The medical device industry is one among many (e.g., aviation, automotive, nuclear power) that places a strong emphasis on risk management and for obvious reasons. Medical device failures and use errors can lead to injury and death. This explains the prevalence of regulations, standards, and guidance pertaining to risk management in the medical device industry.

For our purposes, the most relevant standard is the International Standard Organization's *ISO 14971:2019, Medical devices—Application of risk management to medical devices*. The organization states:

> *This document specifies terminology, principles and a process for risk management of medical devices, including software as a medical device and in vitro diagnostic medical devices. The process described in this document intends to assist manufacturers of medical devices to identify the hazards associated with the medical device, to estimate and evaluate the associated risks, to control these risks, and to monitor the effectiveness of the controls.*

INTERNATIONAL STANDARD	ISO 14971

Third edition 2019-12

ISO 14971 defines the term "risk management" as "the systematic application of management policies, procedures, and practices to the tasks of analyzing, evaluating, controlling, and monitoring risk."

Medical devices — Application of risk management to medical devices

Dispositifs médicaux — Application de la gestion des risques aux dispositifs médicaux

ISO 14971:2019, Medical devices—Application of risk management to medical devices.

A systematic application of risk management calls for mitigating use-related risks, which can be accomplished, in part, by implementing risk-responsive user interface requirements. Risk mitigations come in many forms. *ISO 14971:2019, Medical devices—Application of risk management to medical devices* prioritizes such mitigations:

a) "inherently safe design and manufacture;
b) protective measures in the *medical device* itself or in the manufacturing process;
c) information for *safety* and, where appropriate, training to users."[1]

DOI: 10.1201/9781003029717-8

Such mitigations need to be specified in the form of user interface requirements. However, to identify mitigations, it is necessary to first identify the product's risks. Therefore, while this book is not intended to be a manual for identifying use-related risks, it is an important step in the process of developing user interface requirements, and so we feel compelled to address it. Please bear with us. The balance of this chapter presents steps for developing a use-related risk analysis.

DETERMINE POTENTIAL USE ERRORS

Identifying potential use errors is not complicated. For example, the process might involve inspecting a device in its current form, talking to people who might use the product in some capacity, and performing some desktop analyses to root out potential use errors (a fancy way of saying "a good, old-fashioned brainstorm"). Later, we take a 30,000-foot approach to describing these methods.

TASK ANALYSIS

Generally speaking, task analysis involves listing the discrete steps for using the product in question. The number of steps, sub steps, sub-sub steps, and so on depends on the nature of a user's product interactions and is somewhat a matter of choice. Generally, you can stop when you have identified the significant steps involving perception (P), cognition (C), and action (A). This flavor of task analysis is referred to as a "PCA analysis."

A comprehensive PCA analysis can lead to dozens if not hundreds of perceptions, cognitions, and actions comprising the full set of user interactions with a device.

The following table is a sample of tasks from a blood glucose meter task analysis. As you can see, virtually every P, C, and A can be viewed, quite pessimistically, as an opportunity for use error.

Task (i.e., Perception [P], Cognition [C], or Action [A])	Use Error
Identify expiration date	Overlooked the expiration date
Remember to clean the blood sample site with an alcohol swab	Did not remember to clean the blood sample site
Insert the test strip in the correct orientation	Inserted the test strip upside down
Apply blood to the correct place on the test strip	Applied blood to top of test strip rather than tip of test strip
Read meter's blood glucose measurement	Misread the blood glucose measurement
Remember to annotate blood glucose reading to indicate it was taken before a meal	Did not remember to annotate blood glucose reading

There is sometimes uncertainty if a step should be characterized as a cognitive step or action. The following step, excerpted from earlier, is first characterized as a cognitive step.

C	Remember to clean the blood sample site with an alcohol swab	Did not remember to the clean blood sample site

However, it could just as easily be characterized as an action, as follows.

A	A clean blood sample site with an alcohol swab	Did not clean the blood sample site

Our recommended solution is to include both the cognitive step and action. In doing so, you cover two possible root causes of the ultimate interaction problem of not cleaning the blood sample site. In the first case (C), the device might not make users aware they need to clean the blood sample site. In the second case (A), users might intentionally skip the step because they consider cleaning the site unnecessary.

Task analysis (PCA style or other) is called a bottom-up approach to identifying potential use errors because it calls for (1) identifying discrete steps and then (2) identifying consequences of a use error. The opposite, top-down approach is the nature of a hazard analysis, which we describe next.

HAZARD ANALYSIS

Plainly spoken, a hazard is something that causes harm. Users can be exposed to hazards if they commit a use error. Accordingly, you can identify additional use errors by considering a wide range of hazards, some of which might seem quite unlikely to affect users, and how users could be exposed to those hazards in likely and unlikely scenarios.

The following table is a list of hazards, hazardous situations, and possible consequences that may be present when using certain types of medical devices.

	Hazard and Example Hazardous Situation	Consequence(s)
	Gas Anesthesia gas leaks from a vaporizer or an anesthesia workstation's breathing circuit and is inhaled by the operating room staff.	Explosion, poisoning
	Bacteria, virus Dangerous bacteria (e.g., MRSA)[2] remains on the elevator of a duodenoscope after reprocessing and contacts the next patient.	Infection

	Hazard and Example Hazardous Situation	**Consequence(s)**
	High heat/flame	Overheating, burn
	Heated tip of an electrocautery device touches a surgeon's hand.	
	Chemical	Poisoning, burn
	Toxic cancer treatment fluid spills from an IV bag and onto a nurse's arm.	
	Compressed gas	Gas embolus
	High-pressure oxygen passes through a tube from a wall outlet and into a patient's IV access due to a tubing misconnection.	
	Converging components	Asphyxiation, bruising, hemolysis
	Guardrails on a hospital bed form a "V" that can entrap a patient's head or neck.	
	Electricity	Shock, fire
	Paramedic touches a patient at the moment a defibrillator delivers a cardioverting shock.	
	Moving part	Pinch, crush, laceration, amputation
	Lab technician places hand into a diagnostic device (e.g., to retrieve a sample rack), and it is trapped by a moving component.	
	Radiation	Poisoning, burn
	X-rays strike an unintended part of the body that is not properly shielded by a lead-lined garment.	
	Sharp point (e.g., needle)	Puncture wound
	The caregiver delivers an insulin injection to a family member and then incurs a needlestick injury while attempting to recap the needle.	

It takes a good understanding of a device's intended use environment to identify all of the potential hazards within. Clearly, different types of hazards might be found in diverse use environments, such as the following ones:

- Air ambulance
- Cardiac intensive care unit
- Emergency department
- Operating room
- Physician's exam room
- School nurse's office
- Home (bedroom, bathroom, kitchen, etc.)
- Workplace
- Public setting (e.g., restaurant, theater, sports arena)
- Outdoor setting

Once you have identified the possible hazards of using the product in question, you can identify the potential use errors that could expose a user to such a hazard. You might consider performing a fault tree analysis (or a similar analytical method) for which a review is outside the scope of this book.

Let's consider the example of an insulin pen-injector (remember the recipe in Chapter 2?). In summary, the device enables users to deliver a chosen dose of insulin. The following table illustrates potential hazards associated with using an insulin pen-injector and potential use errors that might expose a user to such harms.

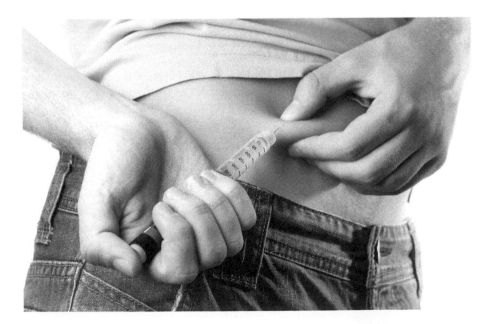

FIGURE 8.1   A person with diabetes uses a pen-injector to administer insulin.

Hazard	Hazardous Scenario	Use Error	Harm
Chemical (insulin)	User misremembers insulin dose prescribed by doctor and cannot get in touch with doctor to confirm the dose	Delivers higher dose of insulin than recommended	Hypoglycemia (low blood glucose level)
Sharp point (needle)	User does not have a sharps container	Disposes of uncovered needle in trash	Needle stick injury, potential infection
Chemical (insulin)	User needs to make room in refrigerator and pushes insulin to the back, where the temperature is colder than recommended for insulin	Stores insulin at incorrect temperature and then injects partially frozen insulin	Ineffective insulin, leading to hyperglycemia (high blood sugar level)

At this point, a fault tress analysis can proceed. We know the harm (e.g., severe hypoglycemia—low blood glucose level—leading to loss of consciousness), which is the consequence of exposure to the hazard (insulin). In this case, you may call it overexposure. So, what led to the exposure and harm? Backtracking, we might think of many use errors that could have led to this outcome (in addition to the one listed earlier), including the following ones:

- Imprecise dial movement led the user to overshoot the target value.
- User misread the dial numbers.
- User inadvertently rotated the dial to a higher setting during handling.

KNOWN PROBLEMS ANALYSIS

There are many proverbs, quotations, and sayings that capture the same basic message, which is to learn from the mistakes you make and those made by others. Perhaps this is the most familiar:

Those who cannot remember the past are condemned to repeat it.[3]

—George Santayana (1863–1952)

Regulators agree that it is imperative to systematically identify issues that have occurred in the past so that they do not happen again. In this context, the process is called a known problems analysis (KPA). At a high level, a KPA identifies use errors committed during the use of products that are similar to the one currently in development. These identified use errors can then be incorporated into a use-related risk analysis to ensure they are mitigated by the new product's design. You can imagine the benefit of a KPA when designing a medical device that has the same functions as many already on the market including its predicate.[4]

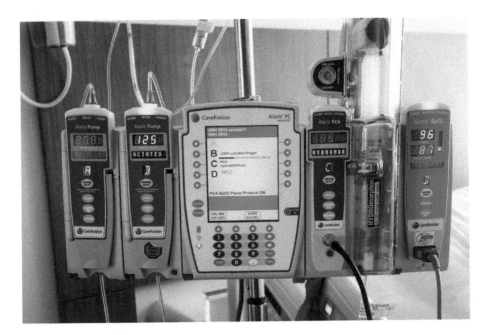

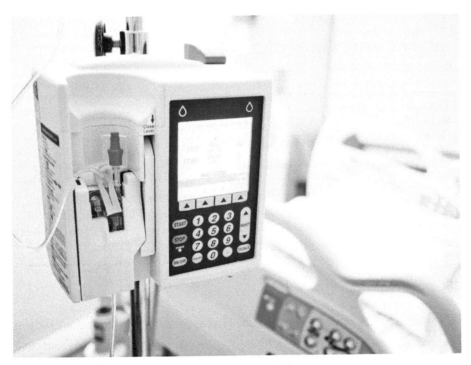

FIGURES 8.2–8.4 Sample intravenous infusion pumps that could be included in a known problems analysis. Note that a pump's inclusion in this table does not necessarily indicate that it was subject to reported problems involving use error.

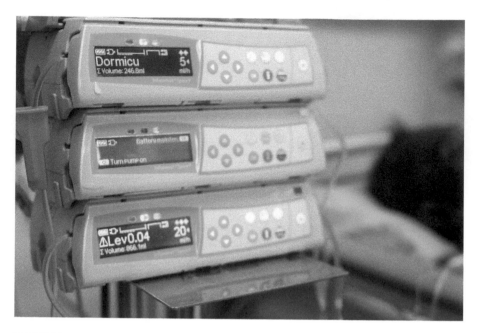

FIGURES 8.2–8.4   (Continued)

The FDA and IEC both expect manufacturers to perform a KPA as part of their risk management efforts:

FDA[5]

6.2 Identification of Known Use-Related Problems

When developing a new device, it is useful to identify use-related problems (if any) that have occurred with devices that are similar to the one under development with regard to use, the user interface or user interactions. When these types of problems are found, they should be considered during the design of the new device's user interface. These devices might have been made by the same manufacturer or by other manufacturers.

IEC:[6]

5.3 Identify Known or Foreseeable Hazards and Hazardous Situations

The manufacturer shall identify known or foreseeable hazards and hazardous situations, which could affect patients, users, or others, related to use of the medical device. This identification shall be conducted as part of a risk analysis performed according to ISO 14971:2007, 4.3 and the first paragraph of ISO 14971:2007, 4.4.

As stated in FDA's footnoted guidance, there are several sources in which to search for known problems with other medical devices. Perhaps the most frequented source is MAUDE[7] (Manufacturer and User Facility Device Experience), which is FDA's online database for device problems (use-related and other).

If applicable, manufacturers can also identify known problems from complaint records they have collected for products that are already on the market. Note that manufacturers with medical devices already on the market are responsible for post-market surveillance (i.e., documenting and responding to [if applicable] product issues reported by users) in some global territories. In the United States, the requirements are spelled out in the Code of Federal Regulations.

CODE OF FEDERAL REGULATIONS. TITLE 21, VOLUME 8, REVISED AS OF APRIL 1, 2019

TITLE 21—FOOD AND DRUGS, CHAPTER I—FOOD AND DRUG ADMINISTRATION, DEPARTMENT OF HEALTH AND HUMAN SERVICES, SUBCHAPTER H—MEDICAL DEVICES, PART 820—QUALITY SYSTEM REGULATION
Subpart M—Records•
Sec. 820.198 Complaint files.

(a) Each manufacturer shall maintain complaint files. Each manufacturer shall establish and maintain procedures for receiving, reviewing, and evaluating complaints by a formally designated unit. Such procedures shall ensure that:
   (1) All complaints are processed in a uniform and timely manner.
   (2) Oral complaints are documented upon receipt.
   (3) Complaints are evaluated to determine whether the complaint represents an event which is required to be reported to FDA under part 803 of this chapter, Medical Device Reporting.

• See Parts b-g in the Code of Federal regulation.

In the European Union, the requirements are spelled out in the Medical Device Regulation:

Regulation (EU) 2017/745 of the European Parliament and of the Council of April 5, 2017 on medical devices (Medical Device Regulation [MDR])
Annex III, Technical Documentation on Post-Market Surveillance, Section 1.1
1.1 The post-market surveillance plan drawn up in accordance with Article 84.

The manufacturer shall prove in a post-market surveillance plan that it complies with the obligation referred to in Article 83.

(a) Post-market surveillance plan shall address the collection and utilization of available information, in particular:
  - information concerning serious incidents, including information from PSURs, and field safety corrective actions;
  - records referring to nonserious incidents and data on any undesirable side effects;
  - information from trend reporting;
  - relevant specialist or technical literature, databases, and/or registers;
  - information, including feedbacks and complaints, provided by users, distributors, and importers; and
  - publicly available information about similar medical devices.
(b) The post-market surveillance plan shall cover at least:
  - a proactive and systematic process to collect any information referred to in point (a). The process shall allow a correct characterization of the performance of the devices and shall also allow a comparison to be made between the device and similar products available on the market.

  . . .

Here is a set of use errors that were derived from a KPA of respiratory devices (e.g., nebulizers, breathing circuits) as an example:

- User did not replace the filter for more than four years, though instructions recommend checking the filter for cleanliness every week and replacing when dirty (MAUDE report 9681154–2016–00005).
- User oriented the device horizontally, resulting in solution from the device leaking onto the patient and causing a burn on the patient's leg (MAUDE report 3004822415–2014–00014).
- User placed the device on a setting lower than what was prescribed by their doctor. As a result, the user suffered hypoxia (MAUDE report 3004822415–2014–00011).
- User did not wait for the water in the device to cool and picked up the device despite its lid being insecure. As a result, the device and lid separated and hot water spilled on the patient (MAUDE report MW5030660).
- Caregivers inadvertently turned off device alarms and, as a result, were unaware that the device was not delivering therapy to the patient. Consequently, the patient died (MAUDE report 9611451–2013–00470).

INTERVIEWS WITH INTENDED USERS

Suffice it to say that you can learn a lot about potential use errors with a particular product by talking to the people who use medical devices, or even better, people who use previous versions of a similar product to the one in development. Such individuals might be clinicians or laypersons, depending on the type of device. For example, if you were developing a hospital bed, you would be well served to speak to nurses, hospital housekeepers, and patients. In the course of such interviews, you might hear about the following use errors:

- Nurse: *"Some people do not take off everything that is on the bed to get the patient's weight, especially when they do not want to disturb the patient."*
- Hospital housekeeper: *"It is difficult to wipe-down the patient beds between patients because the beds are so large, and depending on what else is in the room it can be difficult to access every side of the bed."*
- Patient: *"Oh, I accidentally call a nurse to my room at least once a day because I bump the nurse call button with my elbow."*

INTERVIEWS WITH TRAINERS

Similar to interviewing the intended users of a medical device to identify potential use errors, it can be productive to interview the people who train others to use the device. After all, they are most likely to witness use errors as trainees learn the proper way to use a given medical device. For example, a trainer might recall which steps users tend to forget or the steps that take several attempts to master.

OBSERVATIONS OF PEOPLE USING SIMILAR PRODUCTS

In the course of observing people at work to determine their product needs and preferences, you just might observe a use error, or perhaps more than one. However, do not expect observations of people at work (also called ethnographic research) to uncover too many use errors as compared to those identified by means of a task analysis, for example. This is because use errors occur fairly infrequently in most cases. However, as discussed next, observing people in the context of a benchmark usability test can be productive.

BENCHMARK USABILITY TESTS OF SIMILAR PRODUCTS

A benchmark usability test of similar products (predecessor and competitor products) can contribute substantially to a set of potential use errors because usability tests almost always reveal use errors (a disturbing fact when you are evaluating commonly used medical devices). Such testing also gives you a sense for what mistakes will be more versus less common given the type of user and the use scenarios involved. For example, you might reveal quite a few use errors by having someone engage in the

following use scenarios related to a blood chemistry analyzer, especially if the user
has no prior experience using the given product(s).

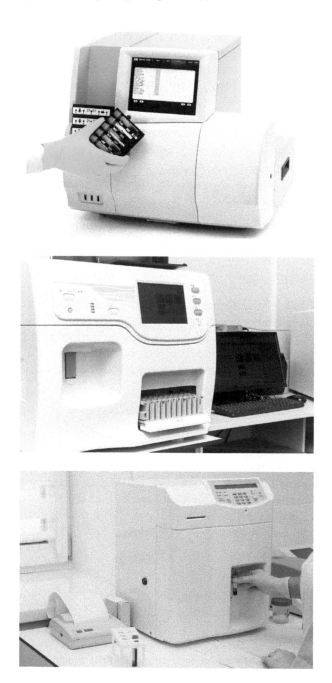

FIGURES 8.5–8.7   Sample blood chemistry analyzers that could be included in a benchmark usability test.

Sample Use Scenarios	Potential Observed Use Errors
Run blood test of sample	User selects incorrect sample type from menu
Interpret test results	User overlooks asterisks indicating abnormal result
Perform routine maintenance	User does not replace the filter

FORMATIVE USABILITY TESTS OF PROTOTYPES

Once you have a working prototype of a medical device, it is a straightforward matter to conduct a formative usability test of it. Similar to the benchmark testing described earlier, it too can reveal potential use errors. If testing occurs early enough in the development process, the use errors can be addressed in a relatively easy, low-cost manner as compared to making changes near the end of development. *Usability Testing of Medical Devices*[8] provides a detailed description of this method.

DESIGN INSPECTION

A thorough user interface design inspection according to established design principles and heuristics, complemented by a dose of pessimism regarding users' performance, might reveal yet more potential use errors. For example, you might find that a pen-injector's pointer is on a different plane than the rotary display of dosage values. This configuration can cause parallax, which means that the pointer can look like it is pointing at different numbers depending on the angle at which the user views the pointer and display.

FIGURES 8.8 AND 8.9   The dose is set to 13 units (left) but, when viewed at a different angle, parallax makes it appear that the dose is set to 12 units (right).

In this case, you would derive the use error of misreading the display when holding the pen-injector in a rotated position.

BRAINSTORMING

Finally, you can also engage in brainstorming to identify potential use errors. There are lots of ways to approach brainstorming that are out of the scope of this book. In summary, the goal is to think in creative and "out of the box" ways that could identify use errors unlikely to arise in the course of other analyses. Often, the result is a list of potential use errors already identified by other means, just a few unique and important ones, and a bunch of absurdities that should be documented but then set aside to focus on legitimate causes of concern. Just be sure you do not misjudge a use error as absurd simply because it is unlikely. The absurd ones should be self-evident, such as:

- Injecting a dose of insulin into the eye.
- Putting a blood glucose meter in a microwave oven.
- Using a nebulizer in a low-gravity environment.

DETERMINE RISK

Once you have identified potential use errors, you will see what you are up against in terms of risk mitigation. In the case of many medical devices, there might be hundreds of potential use errors. However, not all of the use errors will carry the same risk.

As defined in *ISO 14971:2019, Medical devices—Application of risk management to medical devices*, risk is defined as a, "combination of the probability of occurrence of *harm* and the *severity* of that *harm*."

Some use errors will carry a moderate to high degree of risk, suggesting the need for risk mitigation—ideally by design. Some will carry a low degree of risk and not necessarily need to be addressed by design. Later, we summarize the widely accepted means to estimate the risk of use errors.

Rate the Severity of the Potential Harm Caused by the Use Error

According to the FDA, severity ratings carry the greatest weight when it comes to deciding what use-related risks, arising from potential use errors, warrant risk mitigation. The proper mindset, according to some regulators and many HF practitioners, is to explore potential risk mitigations when a use error can lead to even moderate harm, regardless of the probability of the use error occurring.

This risk assessment approach avoids the pitfall of dismissing a risk as acceptably low, simply because you judged it to be highly unlikely (i.e., one in a million). Notably, it is difficult to accurately estimate the probability of a use error occurring. For example, how likely is it that someone will attach a tube to the wrong connector? You would have to conduct an extensive study, confounded by many variables, to develop a reasonable estimate, and such studies for many probability estimates are impractical. Therefore, you are left with making judgment calls about use error probabilities and then discounting them as compared to estimates regarding the severity of potential harms.

Common scales for rating severity and probability are presented in the following table:

Severity	Likelihood
5—Catastrophic Likely to result in death or permanent injury	**5—Frequent** Highly likely to occur in most cases (>80% chance)
4—Critical Potential for severe (and possibly chronic) injury	**4—Probable** Likely to occur in most cases (60%–80% chance)
3—Moderate Potential for moderate, nonchronic injury	**3—Possible** Might occur in approximately half of all cases (40%–60% chance)
2—Minor Potential for minor injury	**2—Remote** Unlikely to occur in most cases (20%–40% chance)
1—Negligible Likely to result in no harm or minor inconvenience	**1—Improbable** Highly unlikely to occur in most cases (<20% chance)

The idea is to rate severity on a scale, which in this case is from 1 (negligible) to 5 (catastrophic). Some analysts choose to use a ten-point scale for the sake of greater granularity in the severity levels. However, ten points of distinction suggest an unrealistic level of precision in view of the confounding variables, such as the type of person incurring the harm.

It almost certainly takes a team that includes a clinician (e.g., physician, nurse, therapist) to converge on appropriate severity ratings. Obviously, we advocate for a HF specialist to be involved, principally because such an individual is likely to be leading the use-related risk analysis and can serve as an advocate for the end user.

RATE THE PROBABILITY OF THE USE ERROR

As mentioned earlier, the probability (sometimes called the "likelihood") of a use error occurring is arguably much more difficult to estimate than the severity of the potential harm caused by it. Some analysts might look toward adverse event reports and complaints to gain some sense for the frequency of certain use errors. However, this approach could be misleading because people are not always aware that they have committed a use error. Even then, use errors are rarely reported. For example, one study of medication errors committed by nurses showed that that the "mean rate of medication errors of nurses was 19.5% while the reporting rate was as low as 1.3% in a three-month period."[9] In this case only, 7% of the errors were reported. We believe that errors involving medical devices are also underreported.

Therefore, there is cause to discount (or even disregard) probability ratings when considering which use errors warrant priority risk mitigation. Nonetheless, the expectation remains that manufacturers will identify a probability estimate to quantify each risk, as described in the following Sidebar.

WHAT IS THE USE OF ESTIMATING USE ERROR LIKELIHOOD?

You might now wonder about the purpose of rating the likelihood of a use error. Well, you can see that frequent and probable use errors would be cause for risk mitigation efforts. Also, likelihood estimates come back into play late in the product development effort, after a HF validation test. During such testing, you might still see a critical use error (i.e., an error that can lead to serious harm) occur. In this case, you would need to determine the root cause(s) of the use error and determine if there were any means to prevent the use error or reduce the associated harm it could cause. If further mitigation is not feasible, then a manufacturer will explain this in its submission to a regulator for approval. Specifically, the manufacturer may assert in its residual risk analysis that the likelihood of the use error is low, and therefore, according to more traditional approaches to risk analysis, further mitigation is not warranted.

CALCULATE THE RISK

It is common practice to quantify risk as a number, which is why traditional failure modes and effects analyses (FMEAs) usually include a column for the risk's rating. Even though one might doubt the legitimacy of quantifying use-related risk, here is the simple formula:

L = Rating of the likelihood (probability) of the use error
S = Rating of the severity of the potential harm arising due to the use error
R = Risk
$R = L \times S$

SAMPLE RISK CALCULATION

Use error: User sets wrong dose on insulin pen-injector dial
Severity: 4 (critical—potential for severe injury)
Likelihood: 4 (probable—likely to occur)
Risk = 4 x 4 = 16

Historically, use-related risks were divided into acceptable and unacceptable sets by finding their place in a chart such as the following one:

		Severity				
		1—Negligible	2—Minor	3—Moderate	4—Critical	5—Catastrophic
	5—Frequent	5	10	15	20	25
	4—Probable	4	8	12	16	20
Likelihood	3—Possible	3	6	9	12	15
	2—Remote	2	4	6	8	10
	1—Improbable	1	2	3	4	5

Traditional parsing of risk values into four levels of concern (e.g., high (red), medium-high (orange), medium-low (yellow), low (green)).

Based on this chart, a risk level of 16 would put a use error in the medium- to high-risk zone and should certainly be subject to risk mitigation efforts. With some variation among manufacturers and their risk mitigation programs, use errors falling in the red and orange zones, and maybe even the yellow zone, should be subject to risk mitigation efforts.

However, in view of the unreliability of probability estimates pertaining to use errors, the medical industry—encouraged by regulators and standards bodies—has called for risk mitigation analyses pertaining to any use error that could cause more than minor harm. A manufacturer might decide that a severity rating of 3 or better would trigger a risk mitigation effort. Similarly, even if the severity rating of a use error is on the low side (1–2), a probability rating of 4–5 might also be considered cause for risk mitigation efforts.

		Severity				
		1—Negligible	2—Minor	3—Moderate	4—Critical	5—Catastrophic
	5—Frequent	5	10	15	20	25
	4—Probable	4	8	12	16	20
Likelihood	3—Possible	3	6	9	12	15
	2—Remote	2	4	6	8	10
	1—Improbable	1	2	3	4	5

Modified parsing of risk values into two levels of concern (e.g., high (red), low (green)).

LIST USE ERRORS WARRANTING MITIGATION AND DEVELOP REQUIREMENTS

The result of the risk analysis process should be a list of use errors warranting mitigation efforts. Restated from the beginning of this chapter, a principal means for mitigating risks is to specify a design that achieves one of the following outcomes:

1. User interface design eliminates the potential for exposure to a hazard. In other words, there is no longer any risk.
2. User interface protects the user from exposure to hazards, perhaps by means of a physical guard or an electronic/computer-based solution that guards against the use error.
3. User interface alerts the user to the presence of a hazard, indicates the consequences of exposure, and directs the user on how to avoid the exposure. In other words, the user interface warns the user to avoid the use error.

So, the task remains to develop the user interface requirements to achieve one of these outcomes regarding each use error deemed to be unacceptable at the onset of the product development effort. This might involve the addition of new user interface requirements to mitigate a newly discovered interaction problem or modification of existing user interface requirements to result in more effective risk reduction.

NOTES

1 International Organization for Standardization. 2019. *14971:2019 Medical Devices—Application of Risk Management to Medical Devices.* Geneva, Switzerland: International Organization for Standardization. https://www.iso.org/standard/72704.html
2 Methicillin-resistant *Staphylococcus aureus.*
3 Santayana, G. 1905. *Reason in Common Sense.* Mineola, NY: Dover Press.
4 A predicate medical device is one that serves as a comparative example of functionality that has already been deemed safe and effective in accordance with FDA's 510(k) process.
5 U.S. Food and Drug Administration. 2016. *Applying Human Factors and Usability Engineering to Medical Devices; Guidance for Industry and Food and Drug Administration Staff.* www.fda.gov/media/80481/download
6 International Organization for Standardization. 2015. *IEC 62366–1:2015 Medical Devices—Part 1: Application of Usability Engineering to Medical Devices.* Geneva, Switzerland: International Electrotechnical Commission. https://www.iec.ch/
7 Manufacturer and User Facility Device Experience. www.accessdata.fda.gov/scripts/cdrh/cfdocs/cfmaude/search.cfm
8 Wiklund, M., Kendler, J., and Strochlic, A. 2015. *Usability Testing of Medical Devices* (2nd ed.). Boca Raton, FL: CRC Press.
9 Bayazidi, S., Zarezadeh, Y., Zamanzadeh, V., et al. 2012. Medication Error Reporting Rate and Its Barriers and Facilitators among Nurses. *Journal of Caring Sciences* 1.4: 231.

9 Designing to Meet User Interface Requirements

CONCEPTUALIZE DESIGN SOLUTIONS

Conceptual design of an overall product and its user interface is usually a task under-taken with a "broad brush." That is to say, it is the time to focus on fulfilling a vision, such as might be documented in a vision statement. It is the time for sketching and block modeling. It is not necessarily the time to focus on meeting discrete require-ments. However, neither is it the time to ignore requirements, particularly because such requirements might actually dictate a conceptual direction.

Here are some user interface requirements that could have a powerful influence on conceptual design.

- The product shall enable use by people who are blind.
- The product shall be maintenance-free.
- The product shall not require any assembly.
- The product shall be waterproof.
- The product shall prevent reuse.
- The product shall be fully recyclable.

As you can see, these user interface requirements promote a certain range of design concepts and preclude others.

In contrast, here are some requirements that might only have bearing later in the design process and are best set aside at the conceptual design stage.

- Displayed characters shall be no less than ten points (10/72nds of an inch).
- There shall be a compact and convenient stowage for the electrical cord (e.g., bracket).
- Audible alarms shall be emitted at a minimum volume of 65 dBA.
- The brake position (engaged/disengaged) shall be visible when viewed from above and on either side of the stretcher.
- Pushbuttons shall provide tactile feedback.

Clearly, these requirements would be relevant at the detailed design stage but not when the goal is to determine a high-level design direction. That said, design sketches can include realistic details that might very well satisfy such specific user interface requirements. Therefore, concept designers are well served to read the user interface design specification in full before beginning their work. However, they should not—as suggested earlier—get bogged down by discrete requirements.

DOI: 10.1201/9781003029717-9

FIGURES 9.1–9.4   Sample early design sketches.

Such sketches are effective at conveying the big idea. Although an industrial designer usually leads the development of sketches, HF specialists and people representing other disciplines are frequent collaborators. Note that these conceptual design sketches are often generated using computer software programs, such as the case of the MRI machine in the following figure. Building a block model that has a minimum of refinement to it is another common form or expression.

FIGURE 9.5   MRI machine drawn using computer software.

FIGURE 9.6   Sample block model of a surgical stapler.

Source: Courtesy of our HFR&D colleague Joe Fegelman.

We admire conceptual design exercises that respect high-level user interface requirements. After all, if the intent is to develop a wearable medical device, there is little or no value in exploring mobile cart concepts. Or is there? Admittedly, there might be value in creative excursions that do not necessarily meet requirements because there really is no harm and there is a possibility of discovering something special, which could lead to a new intent.

So, what are were advising? Should conceptual designers "color within the lines" or venture beyond? The answer probably depends on the type of product, the marketplace, and designers' mindsets. Pressed to give a definitive answer though, we advise developing concepts that fully conform and some that do not and then collect end-user and development team member feedback on the concepts to decide the best path forward. A concept that violates one or more high-level user interface design requirements warrants one of the following adjustments:

- Revise the user interface requirement(s)
- Adapt the concept to conform with the user interface requirement(s)

The automotive design world gives us a good basis of comparison. Car companies routinely design and produce concept cars. These vehicles, only a small fraction of which can be driven, serve as a broad gesture of design direction and give car company stakeholders and sometimes the public (car show attendants) the opportunity to react. Concept cars that receive a strong, positive reaction often move forward into detailed design and engineering and make it to market. However, the marketable product often reflects important changes from the concept car to meet requirements, including those upheld by automobile industry regulators. So, the lesson is for designers to be cognizant of requirements but also indulge their creativity, knowing that the end product will need to conform to the requirements as originally stated or adapted to allow more freedom.

FIGURES 9.7 AND 9.8   Chevy Volt concept car.

Source: NAParish, CC BY-SA 2.0 <https://creativecommons.org/licenses/by-sa/2.0>, via Wikimedia Commons.

FIGURE 9.9   Chevy Volt production car.

Source: Mariordo Mario Roberto Duran Ortiz, CC BY-SA 3.0 <https://creativecommons.org/licenses/by-sa/3.0>, via Wikimedia Commons.

MOST REQUIREMENTS SHOULD ALLOW FOR CREATIVE EXPLORATION

A user interface requirement should not be any more prescriptive than necessary to ensure adherence to an HF-related objective. Stated another way, a user interface requirement should leave the door open—as much as prudent—to many possible solutions. In turn, this opens the door to a wider range of solutions that might satisfy other requirements pertaining to topics such as technical performance, reliability, and cost.

The following requirement pairs illustrate how one may be written in a less or more prescriptive manner that affords more or less design freedom.

Product: Nurse Call System

More Prescriptive

The Nurse Call button should be labeled with a Phone icon, such as the one below:

Less Prescriptive

The Nurse Call button should have a text and symbolic label.

Product: Intermittent Urinary Catheter	
More Prescriptive	**Less Prescriptive**
The package shall include a circular cutout at one end, enabling it to hang on a hook.	The package shall provide the means to hang it on a hook.

Product: Intra-Aortic Balloon Pump	
More Prescriptive	**Less Prescriptive**
The pump shall incorporate a spring-loaded recoiling mechanism to stow the power cord.	The pump shall provide the means to stow the power cord.

Product: Electronic Health Record	
More Prescriptive	**Less Prescriptive**
Patient weight input field borders shall highlight red when entered data are outside the expected range.	The system shall immediately alert users if they enter a patient weight that is outside the expected range.

Product: Glucose Meter's Instructions for Use (IFU)	
More Prescriptive	**Less Prescriptive**
The IFU should include a color graphic to reinforce the textual descriptions of every major step in the blood testing process.	The IFU shall complement text with graphics where they can enhance communication.

ITERATIVELY EVALUATE POTENTIAL DESIGN SOLUTIONS

Guiding concept(s) in hand, product development teams normally progress through multiple design-engineer-test stages, and naturally, this includes a focus on the user interface as well as many other facets of product development. It is a proven way to optimize a design.

As compared to the concept development stage of product development, this is where user interface requirements ranging from high-level to nitty-gritty should have a strong shaping effect. For example, just the way electrical engineers are responsible for meeting a maximum electrical current specification, user interface developers are responsible for meeting a minimum touchscreen target size specification. AAMI HE75 recommends that touch targets are at least 1.5 x 1.5 cm, with 2 cm or more between the touch targets' centers.[1]

Fun fact: Our diagram in Figure 9.11 is actual size, so feel free to use it to confirm whether your touch targets are sufficiently large and spaced.

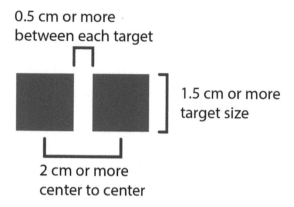

Discussions about effective user interface design and evaluation processes are beyond the scope of this book. However, *Usability Testing of Medical Devices (2nd ed)*[2] contains references to excellent sources of guidance on the topics. Later, we list some of the most common means of evaluation.

DESIGN INSPECTION

This evaluation method calls upon user interface designers and HF specialists (an individual might be both) to critique a user interface in accordance with their judgment, experience, and knowledge of good user interface and HF design principles.

COGNITIVE WALKTHROUGH

This evaluation method calls for representative product users to talk through how they *think* they would interact with a product that is presented in a low-fidelity format, such as static sketches, screens, or physical models.

FORMATIVE USABILITY TEST

This evaluation method calls upon representative product users to engage in use scenarios with a prototype product that might have limited or extensive functional capabilities. Essentially, users take the product for a "test drive" during which they might engage in a continuing dialogue with researchers about their likes and dislikes, or they might work silently.

REVISE DESIGN SOLUTIONS

These types of evaluations give developers a great opportunity to see what aspects of a product are working versus causing interaction problems. Smooth interactions indicate that the pertinent aspects of the product are going in the right direction and the underlying user interface requirements are sound. Conversely, interaction problems usually indicate one of two things:

1. The current design is not meeting the pertinent user interface requirement and needs to be improved.
2. The current design is meeting the pertinent user interface requirement and either (a) the requirement needs modification or (b) the requirement is incomplete and one or more new requirements should be established to address the problem(s).

NOTES

1 Association for the Advancement of Medical Instrumentation. 2009. *ANSI/AAMI HE75–2009: Human Factors Engineering—Design of Medical Devices.* Arlington, VA: Association for the Advancement of Medical Instrumentation.
2 Wiklund, M., Kendler, J., and Strochlic, A. 2015. *Usability Testing of Medical Devices* (2nd ed.). Boca Raton, FL: CRC Press.

10 Conducting Verification Activities

Verification is a conceptually simple matter, even though it might take a lot of work. As discussed in Chapter 3, it involves matching design outputs (i.e., design features) to design inputs (i.e., design requirements). Therefore, to verify the user interface, you are simply matching user interface design requirements to user interface features and documenting the matchups as proof of having "designed the thing right," as FDA has often described verification. Verification documentation includes a verification protocol, which establishes acceptance criteria and the method of verification (e.g., inspection, test, analysis), and a verification report, which details the results of verification activities.

So, what happens if there is a user interface requirement for a physical emergency shutdown button, but the final product does not have one? The answer depends somewhat on a given developer's quality management system and design processes. In principle, the product should be revised to include the button to meet a justified requirement. However, there might be cases where the requirement warrants revision or perhaps even elimination, albeit in a traceable manner.

Developers are almost certain to have an established means to document the verification results, which we would expect to include the following:

- User interface requirements
- User interface feature (general or specific description)
- Image of user interface design feature (optional)

The results might include content such as shown for sample products and associated user interface requirements in the following table.

User Interface Requirements	User Interface Feature	Image
Automated external defibrillator The on/off button shall be color-coded green.	Green on/off button with power symbol indicating on/off.	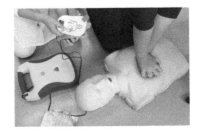

Patient monitor

The monitor shall indicate when it is operating in "Demo" mode.

The word "DEMO" is displayed in the top-left corner of the screen in 60pt font

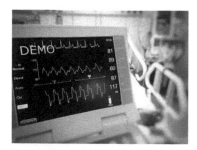

Hospital bed

There shall be an alarm indicating when the patient exits the bed.

A bed exit alarm is placed on the guardrail's external control panel.

Hemodialysis machine

The peristaltic pump's direction of rotation shall be labeled.

Red arrows indicate the direction of pump rotation.

Syringe

A portion of the needle cap shall be at least 1 inch in height or width to enable users with limited dexterity to grip it.

The needle cap has a 1.5-inch, circular handle that enables users to exert more pulling than pinching force.

Endoscope

The insertion tube shall be marked to indicate the length placed inside the patient's body.

The insertion tube's length is marked every 10 cm with white numerals on the black insertion tube.

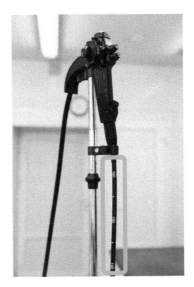

Glucose meter

The test strip shall indicate where to touch a blood droplet.

The gold line printed on the test strip indicates where to touch a blood droplet.

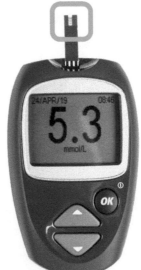

Thermometer

The thermometer shall enable the user to view the temperature reading in no-light use environments.

The display is backlit so that the user can read it in a dark room.

Verification may be done when a design is complete or progressively as design features are locked in. It may be performed by HF specialists or others who can effectively interpret the user interface requirements and judge the adequacy of the solutions—the design features.

Logically, there needs to be a 100% match between the user interface requirements (i.e., design inputs) and the final design features (i.e., design outputs). Otherwise, the design is incomplete or there is a "rogue" design feature that was not subject to specification. The latter case might be of lesser concern because some aspects of the user interface might not warrant specification. An example of this might be the tonal quality of the "click" produced by a button press, although there might be a user interface requirement to produce a "click" sound.

11 Conducting Validation Activities

DESIGN VERIFICATION VERSUS DESIGN VALIDATION VERSUS HUMAN FACTORS VALIDATION

Before we get into the details of validation, let's review some definitions. As mentioned elsewhere, verification answers the question, "Did you design your device right?" and validation answers the question, "Did you design the right device?" In other words, verification provides evidence that your device meets its product requirements (i.e., design inputs). This is largely accomplished through benchtop testing, but it can also be done through certain analyses such as a drawing inspection or FMEA. Validation, on the other hand, confirms that the product meets its user needs. This is done by performing whole system tests with production-equivalent units. The relationship between verification and validation is depicted in Figure 11.1 (published by the FDA).

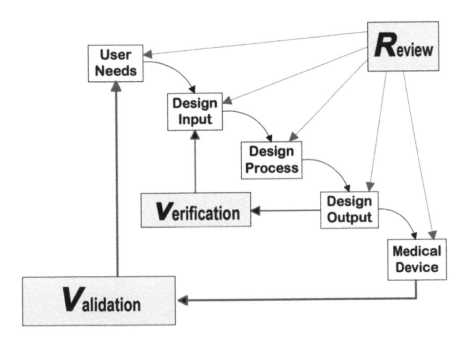

FIGURE 11.1   The relationship between verification and validation.

Source: FDA.

DOI: 10.1201/9781003029717-11

Notably, HF validation is considered part of design validation.[1] So, what makes HF validation distinct from other design validation activities? While design validation's overall goal is to confirm that a device's user needs are met, HF validation specifically evaluates the device's user interface to confirm that the device can be used safely and effectively by the intended users in the intended use environment.

The following table presents a comparison of design verification, design validation, and HF validation.

	Design Verification	Design Validation	HF Validation
Summary	Benchtop testing, analysis, and/or inspection to confirm design outputs meet design input requirements	Simulated/actual use testing (including human factors validation, user interface design validation, and clinical testing) to confirm user needs are met	Evaluation of user interface to confirm intended users can interact with device safely and effectively
Test units	Production units, production-equivalent units, or device subsystems	Production units or production-equivalent units	Production units or production-equivalent units, including all instructional materials, training, and labeling
Timing	During development cycle to end of development cycle	End of development cycle	End of development cycle

Importantly, the three activities can have contrasting outcomes. It is possible, for example, to pass two activities, but fail the third. Here are a few examples.

FIGURE 11.2   Respirator.

Test Type	Acceptance Criteria	Test Outcome
Verification	Respirator enables users to adjust the fit around the nose.	√ Respirator has adjustable nose wire.
HF validation	All participants don and doff respirator correctly.[2]	√ None of the participants commit a use error when donning or doffing respirator.
Design validation	On a scale from 1 (very uncomfortable) to 5 (very comfortable), participants rate the respirator an average of ≥3.5.	⊗ Average rating = 1.5; Participants considered the respirator uncomfortable.

FIGURE 11.3   Dry powder inhaler.

Test Type	Acceptance Criteria	Test Outcome
Verification	Respirator enables users to adjust the fit around the nose.	√ Respirator has adjustable nose wire.
HF validation	All participants correctly report they would clean the inhaler 1/week.	⊗ 5 of 15 participants report they would clean the inhaler 1/month rather than 1/week.
Design validation	On a scale from 1 (very difficult) to 5 (very easy), participants rate the ease of cleaning the inhaler an average of ≥3.5.	√ Average rating = 4.8; Participants considered the inhaler easy to clean.

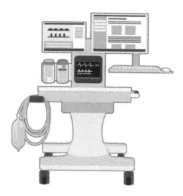

FIGURE 11.4   Anesthesia workstation.

Test Type	Acceptance Criteria	Test Outcome
Verification	Workstation weighs ≤200 lb to enable one person to move it from one use location to another.	⊗ Workstation weighs 225 lb.
HF validation	All participants successfully move the workstation from one room to another.	√ All participants move the workstation from one room to another successfully. No use errors occurred.
Design validation	90% of participants indicate that the workstation's weight is acceptable.	√ 14 of 15 participants (93%) considered the workstation's weight acceptable.

There is no partial credit for going two of three. Rather, successfully passing all three activities—verification, HF validation, and user interface design validation—is essential to arriving at a safe, effective, and satisfying to use product. There is no partial credit.

HUMAN FACTORS VALIDATION

The balance of this chapter focuses on the design validation activities related to the user interface (i.e., HF validation and user interface design validation). There are other design validation activities, such as clinical testing, that are outside the scope of this book.

Let's start with HF validation. The purpose of an HF validation test is to confirm that intended users can interact with the product safely and effectively. HF validation tests involve engaging representative users to complete evaluation activities, including hands-on use scenarios and knowledge tasks. During the evaluation activities, participants perform *critical tasks*. FDA defines critical tasks for medical devices as: "tasks which, if performed incorrectly or not performed at all, could cause serious harm."[3] FDA's definition for drug–device combination products is slightly different:

"tasks which, if performed incorrectly or not performed at all, could cause harm or compromised medical care."[4]

During each evaluation activity, the test moderator (i.e., individual guiding the participant through the test activities) and analyst (i.e., individual collecting and analyzing data) document all observed use errors (i.e., mistakes), close calls, difficulties, and instances of moderator assistance. Then, the test moderator asks open-ended questions to collect participants' subjective assessments of the root cause(s) associated with any such event.

After conducting the test sessions, the test team consolidates and analyzes the performance data with a focus on events with the potential for (serious) harm (and, therefore, related to critical tasks) and then documents the test results in a detailed report.

Here is an isolated example of how a critical task feeds into an HF validation test.

Human Factors Validation Example

Product   Auto-injector

Critical task   Place auto-injector onto injection site at a 90° angle.

Design   The auto-injector's large needle end guides the user to place it at a 90° angle.
mitigation

HF validation   A person with rheumatoid arthritis who has experience using an injection device
test method   participates in a usability test taking place in a usability laboratory. The laboratory is
                    configured to reflect a home environment, similar to the intended use environment.

                    During the test, the participant uses the auto-injector to deliver a simulated
                    injection into an injection pad. During the injection, the test moderator and analyst
                    document test data, such as whether the participant:

- Committed any use errors (e.g., did not place auto-injector onto injection site at a 90° angle)
- Experienced any close calls (e.g., almost did not place auto-injector onto injection site at a 90° angle)
- Experienced any difficulties (e.g., was unsure how to position auto-injector on injection site)
- Required moderator assistant (e.g., to recognize how to position auto-injector on injection site)

After the participant completes the injection task, the test moderator asks several follow-up questions to understand how the task transpired, including debriefing on any observed use errors, close calls, difficulties, and instances of moderator assistance.

Human Factors Validation Example

Acceptance   HF validation usability tests do not have specific acceptance criteria, per se. Rather,
criteria   you should judge the test's success (or failure) based on the outcome of a residual
                    risk analysis. Example outcomes include:

- After assessing all test findings (e.g., use errors) and the associated root causes,
  the level of residual risk is deemed acceptable.
- After assessing all test findings (e.g., use errors) and the associated root causes,
  the level of residual risk is deemed unacceptable. In this case, the manufacturer
  then implements risk mitigations (i.e., design changes) to further reduce the level
  of residual risk.

Note that conducting residual risk analysis requires multidisciplinary perspectives,
including those of medical specialists.

DESIGN VALIDATION

Design validation is the process by which you prove (i.e., validate) that the device
you built works as intended for the end user. This means that manufacturers must
evaluate whether the final product truly meets the defined user needs, as opposed to
only satisfying the preestablished requirements (see Chapter 10).

Recall that user needs often drive the scope of, and details contained within, user
interface requirements. Accordingly, writing user interface requirements and then
implementing a design that meets those requirements are crucial steps in ensuring
that the final product will successfully meet users' needs. That said, if the requirements or the chosen design solution were flawed, then ultimately some user needs
will not be met.

User interface design validation almost always involves conducting a usability
test. Such tests are conducted in a reasonably representative use environment (e.g.,
usability laboratory, simulation center, actual use environment) with test participants
who represent the product's intended users.

During the test sessions, participants engage in test activities, such as performing
hands-on use scenarios and responding to interview and rating questions. Following
the test, the data are analyzed against specific acceptance criteria (defined in the
protocol) to judge whether the user needs are truly met (i.e., validated) and then
documented in a test report.

Here is an isolated example of how a user need feeds into user interface design
validation efforts.

User Interface Design Validation Example

Product   Auto-injector

User Interface Design Validation Example

User need	Users need to know where in the auto-injector the needle is located so that they do not accidentally touch the needle, which can contaminate the needle and/or cause a needle stick injury.
User interface requirement	The device's needle end shall be visually distinguished (e.g., different shape, color) from other parts of the device.
User interface design solution	The device's needle end has a distinct physical appearance, including a large, triangular shape and black color.

The auto-injector's needle end (left) is visually differentiated from the button end (right).

User interface design validation method	A person with rheumatoid arthritis who has experience using an injection device participates in a usability test taking place in a usability laboratory. The laboratory is configured to reflect a home environment, similar to the intended use environment. During the test, the participant uses the auto-injector to deliver a simulated injection into an injection pad. During the injection, the test moderator and analyst document test data, such as whether the participant: • Touched (or almost touched) the auto-injector's needle • Experienced any confusion about the needle's location • Aligned the auto-injector's needle end against the skin and proceeded to administer an injection After the participant completes the injection task, the test moderator asks several follow-up questions to understand how the task went, including, "where is the auto-injector's needle located?"
Acceptance criteria	(1) 0 participants touch the needle during the task (2) 90% of participants correctly identify the needle's location during the debrief PASS = acceptance criteria (1) and (2) are met FAIL = acceptance criteria (1) and/or (2) are not met

If the acceptance criteria are met, then the user need is validated. All that remains is to document the results in the test report.

If you're thinking that HF validation and user interface design validation seem *very* similar, you would be correct. In both cases, test participants perform hands-on tasks with the final product, the test team documents what happens, and then the participant responds to interview questions. The key difference between the two is that HF validation focuses on evaluating whether users can perform critical tasks safely and effectively, whereas user interface design validation focuses on evaluating whether the product meets users' needs.

COMBINING VALIDATION ACTIVITIES

Although HF validation and user interface design validation have different areas of focus (critical tasks versus user needs), there can still be significant overlap. For example, because users need to have safe interactions with the product, many (if not all) critical tasks are likely to align with a user need.

The following table lists example test session activities for HF validation and user interface design validation. As you will see, there is quite a bit of overlap.

Test Activity	HF Validation	UI Design Validation
Session introduction	✓	✓
Review participant background information	✓	✓
Use scenarios focused on critical tasks	✓	✓
Knowledge tasks focused on critical tasks	✓	✓
Debrief (task performance/interaction issues)	✓	
Subjective feedback (safety)	✓	
Additional use scenarios to evaluate user needs		✓
Subjective feedback (e.g., learnability, usability, aesthetics)		✓
Compensation	✓	✓

Even if there are some differences in the test activities and interview questions, there are still considerable similarities in the tests' methods (e.g., recruiting representative users to serve as participants, configuring a representative use environment). An efficiency and cost-minded person might combine the two validation activities.

Combining HF validation and user interface design validation can be an acceptable approach, but we advocate for proceeding with caution and following the best practices described here to increase the chances the process goes smoothly (because no one can *guarantee* validation will go smoothly).

- **Develop a full understanding of HF and user interface design validation activities**. Review the activities required for each validation effort, which will enable you to determine if there is sufficient time in a test session to perform all the activities required for both validation efforts. Sequencing an hour of HF validation activities and 30 minutes of additional user interface design validation activities is reasonable. Conversely, you might have challenges recruiting test participants for a 10-hour test session. Moreover, participants will likely become fatigued if the test session is too long, thereby compromising their ability to realistically perform tasks and adequately respond to interview questions.
- **Complete HF validation session before asking user interface design validation questions**. You must perform HF validation first so that user interface design validation activities do not bias the HF validation's results. The evaluation activities performed as part of HF validation must remain

representative of actual use, and asking user interface design validation questions (e.g., pointing out specific features or aspects of the workflow) can bias the subsequent evaluation activities. Additionally, we are aware of instances where the regulator expressed concern about design validation activities biasing the HF validation results. To avoid any issues, we recommend thinking about it like a relay race. Finish the first leg of the race (HF validation) and then you can handoff to the second leg of the race (user interface design validation).

FIGURE 11.5 Perform HF validation first so that user interface design validation activities do not bias the HF validation's results.

- **Write separate protocols and separate reports.** We suggest keeping the documentation for HF validation and user interface design validation separate. Separate protocols can decrease the likelihood that the two activities become integrated to a biasing extent. The two validation activities have different objectives, as well as distinct reporting needs, so we also recommend reporting on each evaluation in its own report. Separate documentation has the added benefit of clearly articulating to regulators that the manufacturer understands that HF validation and user interface design are distinct.

NOTES

1   U.S. Food and Drug Administration. 2016. *Applying Human Factors and Usability Engineering to Medical Devices; Guidance for Industry and Food and Drug Administration Staff.* www.fda.gov/media/80481/download
2   "Correctly is defined as positioning the respirator over the mouth and nose and adjusting the ear straps such that the respirator creates a seal against the face (i.e., no gaps).
3   FDA's final guidance, Applying Human Factors and Usability Engineering to Medical Devices, issued on February 3, 2016, available on FDA's website.
4   FDA's draft guidance, *Human Factors Studies and Related Clinical Study Considerations in Combination Product Design and Development*, issued in February 2016, available on FDA's website.

12 Sources of User Interface Design Guidance

Herein we list a handful of documents that contain raw and not so raw content (i.e., guidelines, HF data, and design conventions) that you might draw upon to write authoritative user interface requirements.

STANDARDS

AMERICAN NATIONAL STANDARDS INSTITUTE (ANSI)

ANSI Z535 Series

This set of six related standards presents guidance on the design of visual warnings, such as those found on consumer and industrial products (e.g., children's toys and earth moving equipment, respectively). However, they have good applicability to warnings that may appear on medical devices, such as a warning sticker on a hospital bed indicating that brakes should be applied when the bed is not being deliberately moved.

AMERICAN STANDARDS TESTING MATERIALS (ASTM)

ASTM has published over 12,000 standards related to a variety of topics, including those related to product materials, adhesion, electronics, fire and flammability, packaging, to name a few. Additionally, ASTM has published several standards related to medical devices, implants, and medical services (including emergency medical services and related products). While we do not have space to list all of the potentially relevant standards here, consider referencing ASTM's website during your user interface requirement development effort.

ASSOCIATION FOR THE ADVANCEMENT OF MEDICAL INSTRUMENTATION (AAMI)

ANSI/AAMI HE75:2009(R)2013

Human factors engineering—Design of medical devices.

This document contains a large set of user interface design guidelines written specifically for medical devices. While some portions of the document describe HF engineering methods, the majority of the content is design guidance on topics such as displays, controls, workstations, software user interfaces, labeling, and other particularly relevant topics.

DOI: 10.1201/9781003029717-12

AAMI TIR49: 2013

This document is a strong source of design guidelines and criteria for the instructional documents (e.g., IFU, quick reference guides) and training that might be available for medical devices used in a nonclinical use environment (e.g., in a patient's home).

INTERNATIONAL ELECTROTECHNICAL COMMISSION (IEC)

IEC 60601–1:2005+AMD1:2012

> *Medical electrical equipment—Part 1: General requirements for basic safety and essential performance.*

This standard presents a wide array of requirements for medical devices, albeit concentrating on aspects that are normally addressed by mechanical and electrical engineers. Still, some of the requirements have bearing on the design of a medical device's user interface.

IEC 60601–1–8

> *Medical electrical equipment—Part 1–8: General requirements for basic safety and essential performance—Collateral Standard: General requirements, tests, and guidance for alarm systems in medical electrical equipment and medical electrical systems.*

This standard presents many requirements for the design of audible alarms, such as those produced by patient monitors, anesthesia machines, and intravenous infusion pumps; devices chiefly operated by clinicians as opposed to laypersons.

IEC 60601–1–11

> *Medical electrical equipment—Part 1–11: General requirements for basic safety and essential performance—Collateral Standard: Requirements for medical electrical equipment and medical electrical systems used in the home healthcare environment.*

This standard presents many requirements for the design of products intended to be used in nonclinical environments, such as workplaces, public settings, and homes.

IEC 82079–1:2012

> *Preparation of instructions for use—Structuring, content, and presentation— Part 1: General principles and detailed requirements*

This standard outlines principles and requirements for developing IFU across all industries. Although it is not specifically focused on medical devices, it details requirements related to the content, format, and quality of instructional documents that might be a helpful input when developing user interface requirements for medical devices.

INTERNATIONAL ORGANIZATION FOR STANDARDS (ISO)

ISO 11607–1

Packaging for terminally sterilized medical devices—
Part 1: Requirements for materials, sterile barrier systems, and packaging
  systems

This standard specifies requirements and test methods for preformed sterile barrier systems, sterile barrier systems, and packaging systems that are designed to ensure sterility of the packaged medical device or component. Notably, this document does not specify requirements for aseptic packaging nor a comprehensive set of requirements for combination product packaging.

ISO 80369–1:2016

Small-bore connectors for liquids and gases in healthcare applications—
Part 1: General requirements

This standard specifies requirements for small-bore connectors, which convey liquids or gasses in medical devices or accessories, and lists the healthcare fields in which these small-bore connectors are intended to be used. You might find this standard relevant if the medical device you are developing requires tubing connections, for example, and you want to design the device to prevent unintended connections (of course you do!).

ISO 14915 Series

Software ergonomics for multimedia user interfaces—Part 1 (2002): Design
  principles and framework, Part 2 (2003): Multi-media navigation and con-
  trol, and *Part 3 (2002): Media selection and combination*

This series of standards present design principles for multimedia user interfaces, where various types of media might include text, images, audio, video, or other methods of communicating information. Part 1 addresses this topic generally and presents ergonomic considerations for such multimedia user interfaces. Part 2 presents principles related to the content's organization and navigation within the user interface. Part 3 presents recommendations for and guidance on the design and selection of interactive multimedia user interfaces.

U.S. DEPARTMENT OF DEFENSE

MIL-STD-1472

Human Engineering Design Criteria for Military Systems, Equipment, and
  Facilities

This standard describes general human engineering criteria for the design and development of military systems, subsystems, equipment, and facilities. Many of the principles can be applied to other domains, including medical technology. The document is also a source of anthropometric data, although the measurements are primary from military-age males.

NATIONAL INSTITUTE OF STANDARDS AND TECHNOLOGY (NIST)

NIST GCR 15–996

Technical Basis for User Interface Design of Health IT

This document describes HF engineering approaches to developing health information technology, such as electronic health records. Although the document describes methods for evaluating (e.g., usability testing) health IT products, it also presents design principles—many of which are illustrated—which might be more widely applicable to software products.

HUMAN FACTORS AND ERGONOMICS SOCIETY (HFES)

ANSI/HFES 200–2008, Human Factors Engineering of Software User Interfaces

This standard provides requirements and recommendations intended to increase software ease of use, with a keen focus on accessibility and designing software solutions usable by a wide range of users. While this standard is written generally for software solutions, many of the requirements and recommendations are likely to apply to Software as a Medical Device products and medical devices with integrated software user interfaces.

ANSI/HFES 100–2007, Human Factors Engineering of Computer Workstations.

This standard presents specifications for the design of computer workstations, including displays, input devices, and furniture. Although this standard is written generally for computer workstations, some contents might apply to medical devices with integrated graphical user interfaces, such as anesthesia workstations.

U.S. FOOD AND DRUG ADMINISTRATION (FDA)

Guidance on Medical Device Patient Labeling; Final Guidance for Industry and FDA Reviewers (2001)

This standard provides guidance on developing device patient labeling (e.g., IFU, quick reference guides) to ensure that labeling is readable and understandable by users, covering topics such as information organization, presenting warnings and precautions, and general writing principles.

BOOKS

Johnson, J. 2021. *Designing with the mind in mind: simple guide to understanding user interface design guidelines.* Cambridge, MA: Morgan Kaufmann.

This book describes the cognitive and perceptual psychology behind user interface design guidelines, helping readers make decisions on how and when to implement specific design features to increase product usability, in particular, when facing certain design tradeoffs such as competing design principles.

MacDonald, D. 2019. *Practical UI patterns for design systems: fast-track interaction design for a seamless user experience.* New York, NY: Apress.

This book presents various design patterns that, when applied appropriately, can support practitioners with solving various user interface design challenges. After presenting various design patterns, this book presents opportunities for combining patterns and shows readers how to identify variations from patterns that might indicate opportunities for user interface design improvement.

Tidwell, J., Brewer, C., and Valencia, A. 2020. *Designing interfaces: patterns for effective interaction design.* Sebastopol, CA: O'Reilly Media.

This book presents a road map for interaction design, describing activities such as understanding users, organizing software and information so it matches users' mental models, and implementing visual design that supports product usability.

Weinger, M., Wiklund, M., and Gardner-Bonneau, D. 2011. *Handbook of Human Factors in Medical Device Design.* Boca Raton, FL: Taylor & Francis/ CRC Press.

This book is a source for design and usability principles related to medical devices, covering topics such as designing for various use environments, leveraging anthropometric and biomechanical data to ensure designs match the intended users' and their abilities, and conducting usability tests to ensure devices meet users' needs and are safe and effective for use.

Wiklund, M.E., Ansems, K., Aronchick, R., et al. 2019. *Designing for safe use: 100 principles for making products safer.* Boca Raton, FL: CRC Press, Taylor & Francis Group.

This book is a comprehensive source of safety-focused design principles that apply to developing products in any industry, including developing medical devices. Each chapter presents a different design principle, showing how various design features can protect users from potential hazards.

Wiklund, M.E., Dwyer, A., and Davis, E. 2016. *Medical device use error: root cause analysis.* Boca Raton, FL: CRC Press, an imprint of Taylor & Francis Group.

This book serves as guidance to readers on conducting root cause analysis for use errors committed during the course of using a medical device. This book also focuses on design principles that can be applied to the design of medical devices, in the context of identifying user interface design flaws that could be the (or one of the) root cause(s) of use errors.

Wiklund, M. and Wilcox, S. 2005. *Designing Usability into Medical Devices.* Boca Raton, FL: Taylor & Francis/CRC Press.

This book describes HF processes necessary to produce safe, effective, usable, and appealing medical devices, describing hands-on user research methods including usability testing. Additionally, the book shows how designing to meet users' needs leads to improved medical care and exemplifies user-centered design through several case studies.

OTHER SOURCES

Diffrient, N., Tilley, A.R., and Harman, D. 2017. *Humanscale: a portfolio of information.* Chicago, IL: IA Collaborative Ventures, LLC.

This manual serves as a quick reference guide for human anthropometric measurements (among other guidelines), which can be leveraged as an input to designing products that effectively accommodate users' physical size and abilities.

Wogalter, M.S., Conzola, V.C., and Smith-Jackson, T.L. *Research-based guidelines for warning design and evaluation.* Applied Ergonomics 33 (2002) 219–230.

This article presents an overview of empirical literature on warning design guidelines, such as wording, layout, placement, and symbols, and describes approaches to evaluating the effectiveness of such warnings.

WEBSITES

www.usability.gov.

This website is a resource for user experience design best practices and guidelines. The website also provides overviews of a user-centered design process, and methods for ensuring digital content is usable.

www.WC3.org

The World Wide Web Consortium (W3C), which hosts "WC3," a resource for Web standards. While this content is not specifically targeted at medical devices, it is likely to apply to Software as a Medical Device products and medical devices with integrated software user interfaces.

13 Our User Interface Design Tips

Although this book is not intended to be a repository of design recommendations, leveraging strong design principles when developing user interface requirements will bolster the user interface requirements' quality and, in turn, the product's design. For example, the user interface requirement, "The device's text shall have high contrast compared to its background," is fair but does not provide sufficient guidance to the designers choosing the specific text and background colors. Moreover, the requirement as is would not be verifiable. In contrast, the user interface requirement, "The device's text and background shall have a contrast of at least 15:1," provides this more direct guidance that is measurable.

The following sections provide insights and design recommendations that might facilitate your user interface requirements development efforts. The sections include sample user interface requirements. In some cases, the requirements are not sufficiently specific to be verified as is. Rather, we present generalized and high-level requirements; you might consider them a precursor to the final user interface requirements. We expect these general requirements would be revised to ensure they are suitable for a specific device, intended user population, and intended use environment.

CONTROLS

With all of the technological advancement we have seen in recent decades, including the advent of computer-based user interfaces, it is somewhat reassuring that many of the latest products still have some physical controls. Perhaps the most iconic of controls is the pedestrian pushbutton. After all, there really is no substitution for pressing a well-designed pushbutton to get something done quickly. This is true even if the pushbutton takes the virtual form on a touch-screen target.

DOI: 10.1201/9781003029717-13

REACTION TIME

People can react to stimuli and press a pushbutton quite quickly. One website that gives people the opportunity to test their reaction time indicates that, on average, most people can detect a target and click on a computer mouse—a pushbutton of sorts—in less than one-third of a second.

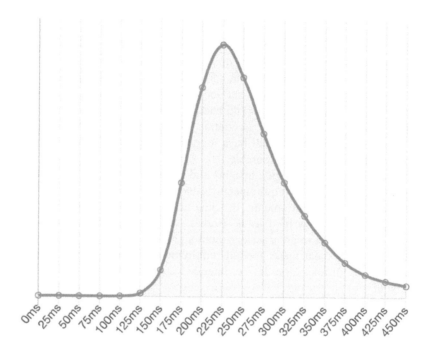

FIGURE 13.1   Plot of the time for individuals to click on a computer mouse when abruptly presented with a visual signal (an indicator turning from red to green).

Source: https://humanbenchmark.com/tests/reactiontime. Disclosure: This website was created by one of the author's family members.

The beauty of a pushbutton is its immediacy, as exemplified by the following examples, each of which is laden with them.

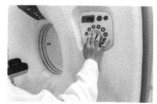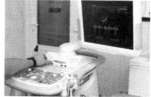

FIGURES 13.2–13.4   From left to right: MRI machine, ultrasound device, and anesthesia machine monitor.

Then, there are many other types of controls that might seem a bit "old school" as compared to digital, virtual ones, but still have a wide array of appropriate applications. The following array shows a small sample of them.

Foot pedal

Hand grip on surgical robot

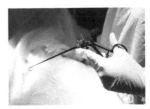

Laparoscopy tool handle

Rotary dial on anesthetic gas vaporizer

Rotary encoder (scroll wheel) on a patient monitor

Sliders on an ultrasound control panel

Trackball on an ultrasound scanner control panel

Rotary knobs on anesthesia gas machine

Hospital bed foot lock

Every control presents the opportunity to optimize user interaction. Here are some general design principles:

- Make the control's purpose and means of actuation self-evident
- Clearly indicate the possible control positions or settings
- Protect controls from unintentional actuation
- Provide positive feedback in response to every control action
- Match control input precision requirements to the users' capabilities
- Provide sufficient feedback to prevent so-called toggle ambiguity

---

TOGGLE AMBIGUITY

Toggle ambiguity occurs when a switch that has two positions does not clearly communicate its current state. For example, a control might switch between the "ON" and "OFF" state, but you cannot tell for sure if a given state is selected or available for selection, such as illustrated later.

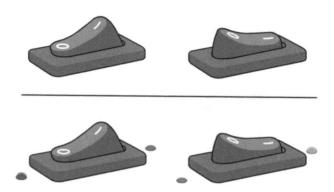

Most people might guess that the indented state is the active state. In other words, the switch on the left is in the "OFF" position and the switch on the right is in the "ON" position. But some people might assume the opposite. Such ambiguities can be resolved by adding an indicator, such as shown on the right side.

Let's now focus now on pushbuttons, noting that many of the design principles that apply to them also apply to other controls. Among the innumerable examples found on medical products, there are "the good, the bad, and the ugly." Indeed, not all pushbuttons are equal, and the worst ones can actually induce use errors. Common flaws include making pushbuttons too small, packing them too close together, presenting too many at once, and designing them in ways that result in very little feedback.

Here are some things to consider when writing user interface requirements pertaining to pushbuttons:

- **Size.** As you would expect, users are going to have an easier time pressing pushbuttons that are sized to match the area of their finger pad. This is why users favor pushbuttons with tops (i.e., caps) that are ≥0.5 inches in diameter. However, smaller pushbuttons can work well if they are spaced sufficiently far apart. Size differences also make pushbuttons more distinct, as illustrated in Figure 13.5.

FIGURE 13.5   Size can help distinguish pushbuttons from each other, as well as convey which pushbuttons are functionally related.

- **Contour.** It often improves usability to make the top of a pushbutton slightly convex (i.e., dished-in); it simply helps center the fingertip in the event of a slightly off-centered touchdown and prevents the fingertip from slipping off and potentially pressing another pushbutton. A flat contour might not be a problem, especially if the pushbuttons are adequately spaced. However, a convex pushbutton (i.e., bulging outward) can be problematic because it will increase the likelihood of the finger sliding off the pushbutton and also might not be considered comfortable to touch.

FIGURE 13.6   Concave pushbuttons, designed to "capture" a fingertip.

- **Spacing.** Basically, you want to space pushbuttons far enough apart to prevent erroneous control inputs. For decades, a center-to-center button spacing of 0.75 inches (1.91 cm) has been the standard of care. This standard dictated the generous spacing of keys on a typewriter and later on a computer keyboard. Recently, the miniaturization of devices and the need for more compact keyboards have led to a reduction in the spacing. Consequently, users have had to press keys more carefully to avoid use errors. The bottom line is that spacing pushbuttons ≥0.75 inches apart is good practice but might not be practical in all cases. When it is not practical, extra attention should be paid to optimizing other pushbutton characteristics that facilitate error-free, comfortable use.

- **Feedback.** You know a good pushbutton when you press it. It gives you feedback in one or more forms that confirms that you struck it properly to actuate the associated function. Traditionally, pushbuttons moved and had force properties that confirmed an actuation. A nice-feeling pushbutton moved perhaps 0.1 inches (0.25 cm) or more, might have produced an increasing amount of resistance force until reaching a "plunge-through" point, and then bottomed out in a definitive manner.

FIGURE 13.7   Pushbutton travel.

Additionally, it might have made a sound that confirmed actuation. Designers who added these kinds of appealing characteristics really knew their "buttonology." Today, many packaging, cost, cleanability, and reliability factors conspire to reduce a pushbutton's positive feedback, making it that much more pleasing when a medical product has great pushbuttons.

- **Shape coding.** Pushbuttons are often rectangular or round but may have other meaningful shapes. Their traditional rectangular and round shapes persist because they are easy to pack into an array and nicely overlay actuators (e.g., dome switches), respectively. However, the traditional shapes might not be as intuitive as other shapes that reinforce the pushbutton's purpose. For example, it is sensible to make pushbuttons that serve to increase and decrease a setting to be up and down arrowheads. As another example, it is sensible to make a pushbutton that stops a process to have a hexagonal shape like a stop sign, at least for use in regions where stop signs are hexagonal.

- **Color coding.** Color coding a pushbutton can elevate or dimmish its conspicuity to match its importance. Color coding can also infer special meaning. For example, a pair of pushbuttons used to turn a pump on and off could be colored green and red, respectively. As another example, a pair of pushbuttons used to make something warmer and colder could be colored red and blue, respectively. Color coding can also reinforce functional relationships. For example, a set of numeric keys could be colored gray to form a single, visual unit of related elements.

FIGURE 13.8   Color-coded pushbuttons.

- **Relationship to display.** When a pushbutton has a closely related display that will show the result of pressing the control, the best location for it is usually below the display or to the right. Placing a pushbutton below a display enables the user to push it without blocking their view of the display except in unusual cases. This means the user can look at the display for immediate feedback in many cases. Placing the pushbutton to the display's right side accommodates right-handed individuals (about 90% of all users) who can press the control with their right hand without blocking their view of the display. Left-handed individuals may also choose to press the control with their right hand or will need to take other steps to keep the associated display in view.

- **Labeling.** Traditionally, the best place for a pushbutton label has been immediately above it. This enables the user to press it without blocking his/her/their view of the label. If control panel space is limited, and in the pursuit of a simpler-looking layout, it also works well to place the label directly onto the pushbutton if an adequately sized (i.e., large enough to be legible) label will fit. It is generally best for a label placed on the pushbutton to contrast with the background material. Molded-in (i.e., embossed) labels can be distinctive, but a contrasting label affords greater legibility, which is likely to benefit all users and especially those with reduced visual acuity.

FIGURES 13.9 AND 13.10   Molded label (left) versus printed label (right).

- **Cleaning.** Membrane-type pushbuttons really make it easy to clean control panels as compared to buttons that stand off the surface and, in some cases, have crevasses surrounding the pushbutton body. Accordingly, it might be wise to use membrane-type pushbuttons when medical product cleanliness is imperative, as illustrated later. In such cases, it is important to design as much feedback into the underlying switches and to perhaps include some surface embossing to reinforce the perception of a physical control. See Chapter 13—Cleaning for more information about cleaning.

FIGURE 13.11   Membrane keypad for the infusion pump.

Source: Presented with permission from the manufacturer, Dyna-Graphics.

- **Durability.** Given the long life of many medical products (e.g., 20–30 years), it is important for pushbuttons to function effectively for the duration. Specifically, they should not become wobbly, lose their haptic qualities, or require multiple tries to trigger a function. Also, the labels should not wear away.

- **Double-press protection.** Pushbuttons can be vulnerable to unintended double presses. Double presses can happen when a switch closes, opens, and closes again, perhaps because the user's finger bounces slightly on a depressed pushbutton. The problem sometimes occurs when a user slides his/her/their finger sideways on a bottomed-out pushbutton and when the user's finger vibrates a bit, perhaps due to essential tremor. The common solution is to add electrical and/or software features that prevent or limit finger bouncing, such as a debounce algorithm to the control software that will ignore any new pushbutton presses for a period of time (perhaps ≤150 milliseconds).

FIGURE 13.12   Debounce algorithms protect against inadvertent double presses.

Now, we will share a few more principles related to other types of controls.

- **Population conventions.** Controls should appear and behave in what we will simply describe as the expected or conventional manner, and therefore, there is a strong need for cultural awareness. This is exemplified by the simple light switch. In the United States, it is conventional for a light switch to be flicked upward to turn the light on and downward to turn the light off. However, people in the United Kingdom as accustomed to the exact opposite behavior. So, you can see the need for care when designing controls, especially if producing a single product for worldwide use so that you are not violating a convention without mitigating the chance of use error in some manner. As another example and an admittedly anecdotal one, people from some regions (e.g., the United States) expect an encoder turned clockwise will move a highlight down a vertical menu while people from other regions (e.g., Finland) expect the opposite behavior.

- **Forces**. The force required to actuate a control should really be no greater than necessary while offering enough resistance to give it a good feel. Controls that move too easily can be perceived as vulnerable to accidental actuation. Also, they might not provide enough resistance to motion to facilitate placing the control in an intermediate position as opposed to overshooting the target setting. For example, people tend to like dials with a somewhat viscous feel or with soft detents that keep them from feeling like they can freewheel. Clearly, the force profile for a foot-operated control (e.g., pedal used to engage a brake) is going to differ from a profile that suits a hand-operated control (e.g., a lever used to release a hospital bed's siderail to place it in a lowered position). In addition to reviewing detailed design guidelines on control actuation forces and human strength data, it might be necessary to test various levels of force with representative users to get them in the right range.

- **Contact surface**. Some controls might benefit from the addition of surface texture and contouring. For example, a lever used to trigger an action could be scalloped to help ensure that the user's fingers do not slip across it.

- **Interlocks**. Certain controls might warrant the protection from unintended and erroneous actuation that comes from interlocks. Simply stated, an interlock is some means to prevent a single control action from producing an effect, such as activating a pump, energizing a laser, or transecting and stapling a vessel. Interlocks are usually implemented when an unintended or erroneous action could be dangerous or damaging. An interlock may require the user to perform one of the following sample actions.
  - Click "OK" on a confirmation dialog box displayed on a screen
  - Press and hold a pushbutton for 5 seconds while viewing a countdown display
  - Simultaneously actuate a second control
  - Perform two control actions in series
  - Require the device to be properly configured (e.g., staples loaded into a surgical stapler) before it can be actuated

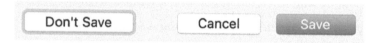

FIGURE 13.13   Confirmation dialog serves as an interlock, requiring the user to confirm a control action.

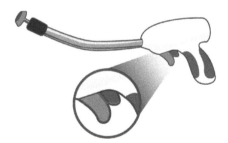

FIGURE 13.14   Surgical stapler has a built-in safety (in red) to prevent inadvertent actuation.

SAMPLE USER INTERFACE REQUIREMENTS FOR CONTROLS

- Each pushbutton shall perform a single purpose.
- Pushbuttons shall have at least 2.5 mm of travel to provide tactile feedback.
- Pushbuttons shall be ≥0.5 inches (1.27 cm) across in all directions, regardless of shape.
- Pushbuttons centers shall be spaced ≥0.75 inches (1.9 cm) apart.
- Pushbuttons shall be either flat or concave, but not convex.
- Pushbuttons shall have debounce protection.
- Pushbutton labels shall contrast with their backgrounds (contrast ratio shall be ≥7:1).

DISPLAYS

Presently, a large proportion of information is conveyed by computer-based displays built into or connected to medical products. This includes big displays, such as those found on diagnostic and therapeutic workstations, but also little ones, such as those found on over-the-counter devices like a noninvasive blood pressure monitor and digital thermometer. Still, there are plenty of "classic" displays to be found, such as the pressure meters connected to oxygen bottles, the dose counters built into drug injection devices, and the power level readouts found on electrosurgical devices (i.e., generators). As a result, there is a lot to be said about designing effective displays; a volume of guidance that goes well beyond the scope of this book.

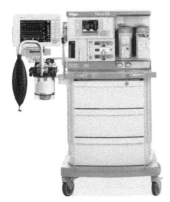

FIGURES 13.15 AND 13.16   Anesthesia workstation and a digital thermometer, which each incorporate a computer-based display.

FIGURES 13.17 AND 13.18   Oxygen tanks with pressure gauges and insulin pen-injector, each featuring more traditional displays.

In view of the caveat earlier, here is just a sample of the kind of principles you should consider when designing various types of displays.

- **Content legibility.** Good legibility arises from many of the same information display properties, whether you are talking about words printed on paper or a control panel or information presented on some kind of

---

IRRADIATION

Irradiation occurs when you place white characters on a black background.

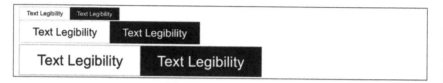

The characters appear to glow to some extent, which can make it more difficult to discriminate small details (e.g., the white space inside a lower case "a"). German scientist Hermann von Helmholtz coined the term "irradiation illusion" in the 1860s to describe how a light area appears larger than an identically sized dark area.

Legibility also arises from using relatively simple fonts and sticking with moderate line thickness and aspect ratios. Content size is also important, owing to the limits on human visual acuity.

Rules of thumb call for the height of alphanumeric characters to subtend 12 arc minutes at a minimum, with the preferred visual angle between 20 and 24 arc minutes. For example, for text that is likely to be read at 3 feet, character height should be a minimum of 0.126″ (0.32 cm, 9pt font) and ideally would be between 0.209″ (0.53 cm, 15pt font) and 0.251″ (0.64 cm, 18pt font).[1]

Example: You can help ensure that the signal word WARNING on a warning sign is legible and draws attention by making it relatively large. Following the recommendations set forth in AAMI HE75, you might choose to make the letters about 0.25″ tall (0.64 cm) tall if you expect the sign or sticker to do its job at a distance of 3 feet (91.4 cm).

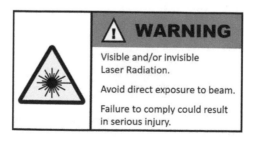

FIGURE 13.19   Warning sign example.

computer-based display. For starters, legibility depends on high contrast between the content (e.g., alphanumeric characters, symbols, demarcation lines) and their background (i.e., field). As noted elsewhere in this book, black content placed against a white background has the highest contrast, without there being any irradiation.

- **Placement.** As obvious as this might be, you need to place displays where they will be visible based on the task at hand. That is, the display should be directly in view from the user's position when she or he is performing a task that depends on reading the display. What we mean by "directly in view" is contextual. In some cases, a display might need to be straight ahead. In other cases, a display might need to be within a 30-degree cone of vision. Sometimes, it might be acceptable for the user to turn to see it, but not reposition her or his entire body. When there is a strong display-to-control relationship, the display goes best above the control so that the reaching hand does not block the user's view of the display.

- **Precision.** Displays should present information at the "just right" level of precision. For example, it is quite sufficient—medically speaking—to present blood pressure values using whole numbers (i.e., integers). This is why you normally see blood pressures displayed as follows:

    There would be no clinical value in any greater precision, which is why you do not see pressures displayed as precisely as 106.2 mmHg over 64.8 mmHg. Conversely, there might be a need for greater precision when weighing a newborn baby because fractions of ounce might be medically relevant to growth patterns and weight-based drug doses.

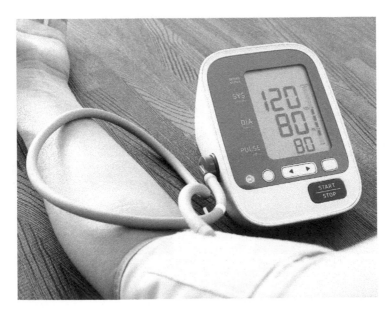

FIGURE 13.20   Blood pressure readout in whole numbers.

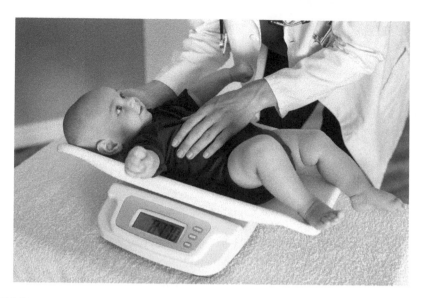

FIGURE 13.21   Scale displays baby's weight to single decimal point accuracy.

- **Decimal points.** Small decimal points can pose a serious hazard when they go unnoticed. Medical device users can mistake a number like 8.7 to be 87; tenfold the actual number. Such use errors have actually led to fatal adverse events. Therefore, it helps make decimal points larger than the conventional dot on the lower case "i," for example. A decimal point formed by displaying a single pixel might be too small, depending on the display's resolution (i.e., pixels per inch). LCD displays facilitate formatting decimal points to be distinctive; for example, the decimal point can be presented with rounded edges.

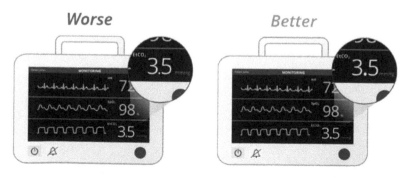

FIGURE 13.22   Comparison of a small versus larger decimal point.

Source: Designing for Safe Use (2019), CRC Press.

- **Use of color.** An old rule of thumb calls for designing display in grayscale and then coloring only certain elements when it adds information. This approach helps ensure that color is used purposefully and not simply to decorate. While it might not be necessary to design in all gray, there is still wisdom in displays presenting most information in an unadorned manner and then using color to make important information stand out, to reinforce functional relationships, and to communicate special meaning. This design approach is illustrated in Figure 13.23.

  Limiting the use of color on a software user interface to 5–7 colors helps avoid a garish appearance and sustains the ability of carefully assigned colors to convey meaning.

FIGURE 13.23 CT scanner control display, with principally gray controls and one red button, indicating that button's functionality and high importance.

- **Units of measure.** Units of measure should be displayed in close proximity to the associated measurement. In some cases, it may be acceptable to allow users to hide the units of measure displayed on a computer screen when they are familiar with those units of measure, the data can only be expressed one way, and there is no chance of a critical error. When there is a chance of critical error, the units of measure should be presented for the sake of safety and usability. This is particularly true when a given product can display measurements in English System (also called Imperial) units or Metric System units, and/or when a drug concentration can be based on volume (milliliters) versus weight (milligrams). Limited display space often leads designers to show the units of measure in very small text that makes them inconspicuous, sometimes bordering on illegible. Therefore, make the units of measure larger when they are quite important to acquiring the correct information.

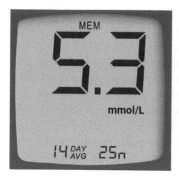 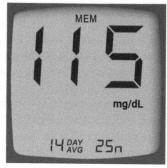

FIGURE 13.24 Similar glucose meters display glucose readings using different units (mg/dL versus mmol/L) to satisfy users in different markets.

- **Screen layout.** There is certainly both an art and science to software user interface design. The artistic part usually focuses on making a user interface look pleasing, or at least to have the intended impact and reinforce a company's brand. You might want to write some user interface requirements that address artistic concerns. Not all requirements have to focus on safety, for example. That said, we are mostly concerned about the science part of designing a software user interface, and screen layout is an important element. A good screen layout should result from keen attention to these goals (among numerous others):
  - Reinforce functional relationships among on-screen elements by placing them in related groups.
  - Arrange elements in task-oriented order (mirroring a sequence of steps), working from left-to-right and top-to-bottom when designing to suit user populations that read documents in this order. Adapt the order to the cultural expectations of the targeted user population.
  - Avoid placing too much information on a single screen unless all of it needs to be viewed together to support the task(s) at hand. Conversely, avoid placing too little information on a single screen because this will require the users to do more paging and/or scrolling.
  - Place information in consistent locations. For example, you should place the time and date in the same location on various screens, rather than many different locations. This facilitates rapid information acquisition. Place controls in consistent location for the same reason.
  - Use blank space judiciously to separate related groups of information and to limit content density, noting that screens packed with too much content can look intimidating and unattractive and can increase the time it takes to locate content of interest.

- **Differentiate information and controls.** It is important for users to be able to readily differentiate information from controls. This becomes difficult when the two types of screen content look similar. A conventional way to distinguish the content is to make only the controls look somewhat three-dimensional by giving them highlighted and lowlighted edges, while designing text and other static features to appear "flat," for example.

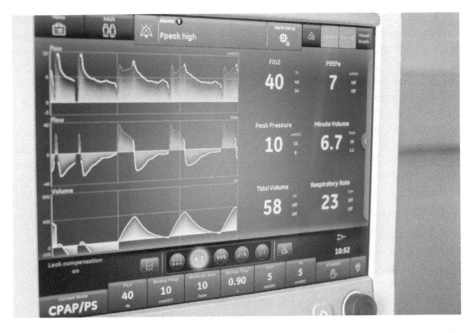

FIGURE 13.25 This patient monitor's active buttons and clickable fields feature a color gradient and somewhat rounded edges, whereas static information is presented as "flat" in the background.

Sample User Interface Requirements for Displays

- Each screen shall have a meaningful title that indicates the screen's primary function.
- Screens shall use ≤7 colors.
- Text shall be sans serif to ensure legibility.
- Text shall be ≥12 points to ensure legibility.
- Displays shall automatically dim or turn off after ≥5 minutes of inactivity.
- The units of measure shall be displayed on the right side or below the value.
- Heart rate values shall be displayed as integers.
- The 12-hour time format (e.g., 5:30 PM) shall display the "AM" and "PM" at the same size as the time numerals.

ALARMS

Alarms can save lives by drawing attention to a hazardous situation. For example, patient monitors can be set to produce a visual and audible alarm if a patient is experiencing a life-threatening arrhythmia. Alarms can also draw attention to other important events that might not be life-threatening but could interfere with effective patient care. This is why alarms usually ascribe to a hierarchy of importance, ranging from (1) high, to (2) low, to (3) advisory; reinforced by a matching set of distinguishing design features.

Visual alarms are usually differentiated by signal word (e.g., High, Low, Notice), text background color (e.g., red, yellow, blue), size, placement on a screen, and perhaps flash rate. Audible alarms are usually differentiated by the number of tones comprising a single "burst," whether or not a burst (i.e., a series of sounds comprising a complete signal) repeats, and sometimes their volume. And, in a manner familiar to many, alarms may also be communicated through tactile cues; the way a cellphone may vibrate while in silent mode to attract attention.

Despite all the good they do, alarms can also cause problems when users habituate to them. This is often the case in clinical environments where you may hear a cacophony of alarms emitted by what might be hundreds of individual medical devices. The problem is summarized by the following findings:

> "Staff at a 15-bed unit at Johns Hopkins Hospital documented an average of 942 alarms each day—or 1 every 90 seconds (no wonder that their new high-tech hospital prioritized noise reduction to improve patient comfort). While some of these alarms are critical, many are false alarms or alarms to indicate the status of the machine itself and not the patient. Such overstimulation can potentially lead to desensitization, a phenomenon known as 'alarm fatigue.' It's a workflow problem with dire consequences: alarm fatigue resulted in over 200 deaths nationwide in the second half of the 2000s."[2]

Consequently, medical product developers are challenged to design alarms that effectively draw the user's attention and, in many cases, give them an immediate sense of the problem at hand, but that also protect users from false alarms and an excessive number of alarms—a tall order.

Here are some principles pertaining to the design of effective alarms:

- **Hierarchy.** Assuming that not all alarms are of equal importance, they should ascribe to a hierarchy of importance. Many medical products bifurcate alarms into two categories: high priority and low priority. However, in some cases, it might be appropriate to add a middle category: medium priority. The different priorities may be expressed using one or more of the characteristics covered by the following principles.

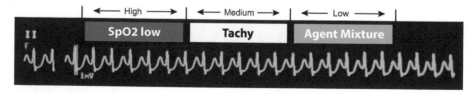

FIGURE 13.26 Annotated screenshot from a patient monitor clinical reference guide reflects the hierarchy of high (red) and low (yellow) alarms along with advisories (blue).

- **Color coding.** Various standards, such as AAMI HE75, call for high-priority alarms to be red, in the form of either a red flashing light or white text on a red background, for example. Low-priority alarms are usually depicted using yellow. If there is a medium priority alarm, the associated color is typically orange. This color-coding scheme aligns well with the scheme that is conventionally used in safety signs.[3]

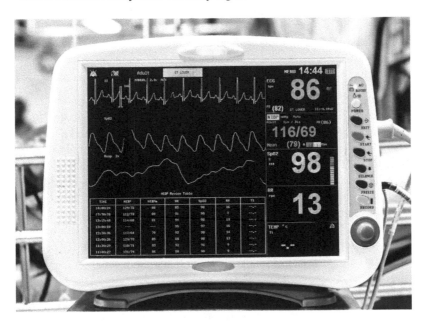

FIGURE 13.27   Even if the user cannot read the alarm details from a distance, the patient monitor is displaying an alarm condition in the form of the yellow rectangle.

- **Symbols.** Controls (e.g., physical buttons, touchscreen buttons) associated with visual alarms are normally labeled using standard symbols, which might be augmented by text labels. The IEC's symbols[4] are in widespread use, although some manufacturers still choose to use their own symbols. In the latter case, a manufacturer would carry the burden of proof that users properly interpret the symbols. The IEC's symbols achieve the worthy goal of standardization, enabling users to learn the meaning of one set of symbols and apply that knowledge across many different medical products. The only issue with the IEC symbols is that some users have found it difficult to initially intuit their meaning and then remember it.

FIGURE 13.28   IEC 61601–1–8 symbols that represent (left-to-right): Audio paused, audio off, alarm, alarm reset, alarm off, and alarm paused.

- **Smart alarms.** An increasing proportion of medical products incorporate "smart alarms," or a smart alarm system. For starters, such systems help users—addressing the alarm fatigue problem—by filtering out nuisance alarms. For example, a smart alarm system that "sees" an alarm condition may wait for corroborating inputs before calling attention to the situation. That is, the system may initially assume it is seeing what most likely is an artifact rather than a real hazard. Smart alarms also alert users to hazardous situations before they are likely to become acute. The alarm system might see a steadily deteriorating medical condition, such as a steadily increasing blood pressure reading, and sound an alarm early enough that a clinician can take action before the patient's blood pressure rises into the high danger zone.

- **Sound.** IEC 60601–1–8 provides a great deal of guidance on how to design a medical product's audible alarm(s). It speaks to the need for audible signals to have appropriate frequency, sound pattern (i.e., melody and harmonics), burst rate, and other aural characteristics to make them attention-getting and distinct. Ultimately, it is essential for an audible alarm to be noticeable against the ambient noise by being sufficiently loud and distinct. For example, a device that needs to stand out in a crowd of people talking would need to be many decibels above the crowd noise (e.g., perhaps rising from a low of 65 dB to a high of 85 dB, which is equivalent to a kitchen blender at close range) and a higher frequency than human voices. The IEC standard provides guidance on how to accomplish these goals and more.

- **Flashing (i.e., blinking).** Sometimes, it is not enough for a visual signal to appear, or change color, or perhaps become larger because the information appears static at a glance. This is when a flashing signal might be useful, at least to draw attention to critical alarms. But, the problem with a flashing signal with a 50% duty cycle (i.e., the on and off times are equal) is that it might have flashed off at the very moment someone glances toward and then away from it. For this reason, it sometimes is best to use a duty cycle biased toward the "on" state. For example, signal detection is likely to improve if a signal is "on" 3/4ths of the time and "off" 1/4th of the time over a total duration of 1 second, which equates to a 75% duty cycle.

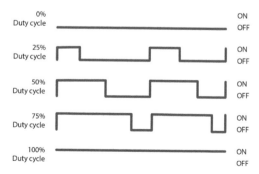

FIGURE 13.29   Various duty cycles.

- **Latching and nonlatching.** A latching alarm is one that will continue unabated until the user attends to it, perhaps by pausing the audio before addressing the alarm condition. A nonlatching alarm will stop whenever the alarm condition goes away. In the latter case, an alarm condition might appear and disappear without the user ever being aware, which could be problematic. Therefore, it is important to decide which alarms should be latching and nonlatching so users do not lose important clinical information. For example, you would want to make an alarm linked to a dangerous heart arrhythmia alarm (e.g., asystole) a latching alarm.

- **Activation.** A history of adverse events suggests that alarm systems should remain active at all times instead of having controls that enable users to turn them off. Although enabling users to turn off alarms might help alleviate alarm fatigue, it defeats the whole purpose of providing protective alarms and, therefore, is ill-advised. Alarms are usually an essential risk mitigation.

- **Limits.** Depending on the medical device, it might be best to establish alarm limits for all users or enable users to set the limits. Usually, clinicians appreciate a device that has "factory default" limits or preset limits established by their healthcare institution. That said, clinicians also value having the option to change limits to suit the conditions of patient care. When a device enables limit adjustments, it is usually over a clinically appropriate range. For example, it might be appropriate to adjust an upper heart rate alarm from 130 to 150 beats/minute to account for a patient experiencing atrial fibrillation with a rapid ventricular response whose resting heart rate is 135 beats/minute. Adjusting the alarm limit in this case will reduce the number of alarms produced due to exceeding the upper limit. Figure 13.30 is an example visualization of lower and upper alarm limits. Anything in the red range is high priority, while the yellow is medium priority.

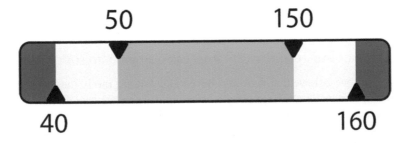

FIGURE 13.30   Alarm limit adjustment.

- **Logging.** It can be helpful if not essential for a medical device to maintain a log of all alarms, be they latching or nonlatching. Clinicians may look at a log to determine if there is an informative pattern that could help them make medical decisions. A log might also be necessary to meet patient care documentation requirements.

SAMPLE USER INTERFACE REQUIREMENTS FOR ALARMS

Visual Alarms

- Alarms notifying users of hazardous situations that, if not avoided, will result in death or serious injury shall be color-coded red.
- Alarms notifying users of hazardous situations that, if not avoided, could result in death or serious injury shall be color-coded orange.
- Alarms notifying users of hazardous situations that, if not avoided, could result in minor or moderate injury shall be color-coded yellow.
- All warning symbols used in the user interface shall align with ISO standards.[5]

Audible Alarms

- Alarm sound levels shall be adjustable to accommodate various auditory conditions.
- Alarms reflecting hazardous situations that, if not avoided, could or will result in death or serious injury shall be latching.
- Alarms reflecting hazardous situations that, if not avoided, could result in minor or moderate harm should be latching.
- The system shall maintain a history of all system alarms from the past 60 days.
- The alarm limits shall be adjustable.

WARNINGS

First, we offer a philosophical statement about warnings that is a rebuttal to claims that warnings are useless. We believe the following:

SOME WARNINGS HELP SOME PEOPLE IN SOME CIRCUMSTANCES TO SOME DEGREE

That is to say, we believe that warnings have value but are rarely the complete solution when it comes to protecting people from exposure to hazards. Consider the fact that some people might not even notice or read the warnings that come with their power tools, but that warnings about sharks and alligators in the water will persuade most, if not all, bathers to stay high and dry.

Ultimately, a warning's effectiveness depends largely on the viewer's risk perception, which, in turn, depends on a person's understanding of (1) the hazard and harms that

FIGURE 13.31   Shark sighted warning.

could arise from exposure; (2) past experience dealing with or learning about hazardous situations; and (3) the design and placement of the warning. From this point forward, we will focus on the last point, which connects to the task of defining user interface requirements.

Here are some fundamental design goals to consider ahead of creating a warning:

- Draw attention to the warning because it cannot communicate anything to anybody if it goes unnoticed.
- Convey the severity of the hazard using the right signal word so that people give it an appropriate level of attention. Yes, you can assume that people might give more attention to a sign that says WARNING instead of CAUTION. Likewise, they might give even more attention to a sign that says DANGER.[6]
- Communicate the potential harm that may arise from exposure to the hazard. This can convince people to take precautions they might not take otherwise.
- Communicate with people who might not speak the native language or who have low literacy.
- Make the warning content memorable. This may be accomplished by adding a graphic that reinforces the text, for example.
- Size the warning to enable effective placement. For instance, it might be necessary to design a small, slender warning so that it can be placed directly in the view of a device's user.

FIGURE 13.32   Sample warning sign.

Warning designers frequently turn to a collection of standards issued by the American National Standards Institute (ANSI) Z535 series for guidance, including the following standards:

- ANSI Z535.1—Safety Colors
- ANSI Z535.2—Environmental Facility and Safety Signs
- ANSI Z535.3—Criteria for Safety Symbols
- ANSI Z535.4—Product Safety Signs and Labels
- ANSI Z535.5—Safety Tags and Barricade Tapes (for Temporary Hazards)
- ANSI Z535.6—Product Safety Information in Product Manuals, Instructions, and Other Collateral Materials

Note that this guidance is suited to a broad set of products ranging from step ladders to earth-moving equipment. However, they have general applicability to the warnings placed on medical devices. Note that they are less applicable to warnings appearing on software user interfaces.

Here are a couple of warnings that generally ascribe to the ASNI guidance.

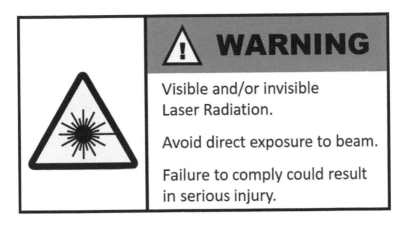

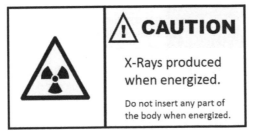

FIGURES 13.33 AND 13.34   Sample warning and caution signs.

SAMPLE USER INTERFACE REQUIREMENTS FOR WARNINGS

Format

- Warning content shall ascribe to one of the following formats (signal word choice to be determined) based on its size and where it will be placed.
- Warning text descriptions shall have a light appearance, produced by black text on a white background.

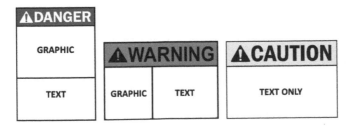

FIGURE 13.35   Danger, Warning, and Caution formats.

Signal Words

- Warnings shall include a signal symbol plus signal word pair to indicate the severity of the potential harm that could result from exposure to a given hazard. Signal words include:

- Warnings shall include a graphic that depicts either (a) the hazardous situation to be avoided or (b) the precautionary behavior being encouraged. In the first case, the graphic should convey a prohibitive message (e.g., "Do not kink tube"), and in the second case, it would convey a permissive message (e.g., "Wear protective eyewear").

- Warnings shall include a statement of the consequence of noncompliance (i.e., what can happen in the event of exposure to a hazard).

- Warnings shall state how to avoid a hazard (e.g., "Do not stare into beam").

- Warnings shall be protected against undue wear and tear over the product's expected service life.

DANGER indicates a hazardous situation which, if not avoided, will result in death or serious injury.

WARNING indicates a hazardous situation which, if not avoided, could result in death or serious injury.

CAUTION indicates a hazardous situation which, if not avoided, could result in minor or moderate injury.

FIGURE 13.36   Definitions for Danger, Warning, and Caution.

Text

- Fonts used in warnings shall be sans serif (i.e., plain characters with no graphical flourishes).

- The hazard shall be presented using bold text.

- Warnings shall be produced in the intended users' native language.

- Warning text shall communicate effectively to people with a fifth-grade reading level, as measured using the Flesch-Kincaid readability formula.

Graphics

- Graphics shall be computer-generated line drawings (i.e., no photos or hand drawings).

- A drawn object's features shall be simply outlined (i.e., should not have more than three different line weights).

Placement on Product

- Warnings shall be placed where they are highly likely to be in view when the user is performing closely associated tasks (i.e., on device's front or side panels).
- Multiple warnings shall be placed on the product when necessary to ensure that they are visible to the user at the right moments during one or more tasks.

CONNECTORS AND CONNECTIONS

A large proportion of medical devices incorporate one or more components that connect together in a physical manner. Here are some examples:

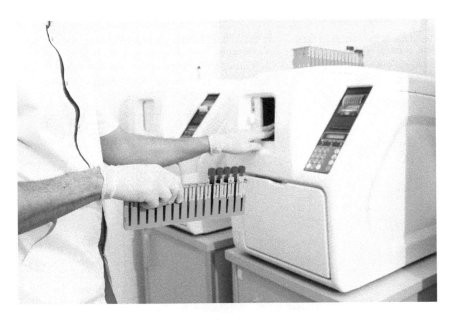

FIGURE 13.37   Table-top analyzer.

Table-Top Analyzer	
Connections	**Comments**
Sample rack that inserts into a table-top analyzer that tests human serum to detect certain viruses.	It is important for the rack to slide easily, in the correct orientation, into the slot and make a secure connection. A connection problem could delay a test result and possibly waste a precious sample.

FIGURE 13.38 Anesthesia machine gas cylinder.

Anesthesia Machine	
Connections	**Comments**
Gas-carrying tubes that connect to an anesthesia machine.	It is important for the tubes to remain securely attached to the gas ports to deliver gas to the patient. A connection problem could cause gas leakage into the care environment.

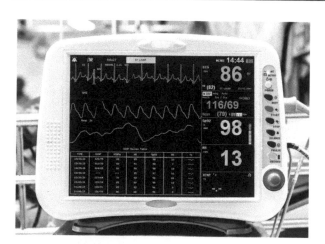

FIGURE 13.39 Patient monitor.

Patient Monitor	
Connections	**Comments**
Electrode lead that plugs into an electrical port.	It is important for the leads to connect securely to the correct port so that the monitor can display, without interruption, the measured values correctly. Connection problems could leave clinicians without essential information to determine proper patient care.

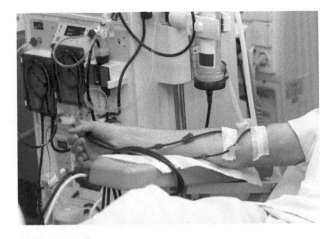

FIGURE 13.40   Hemodialysis machine.

Hemodialysis Machine	
Connections	**Comments**
Tubes that extend from a hemodialysis machine and connect to a patient's vascular access site.	The tubes must enable the user to make a proper connection to the patient's arm to ensure proper blood flow. Connection problems could lead to hypovolemia or hemolysis.

FIGURE 13.41   Insulin pump.

Insulin Pump	
Connections	**Comments**
Tubes that connect an insulin pump to an infusion set.	It is important for the insulin line to be properly and securely connected to the pump and the infusion set, which, in turn, delivers insulin to the patient at the programmed rate. Leaking insulin could lead to hypoglycemia.

FIGURE 13.42   Inhaler with spacer.

	Inhaler
Connections	**Comments**
An inhaler's mouthpiece connects to a spacer and facemask.	The inhaler must be fully inserted into the spacer to ensure that the sprayed medication reaches the patient. An incomplete or insecure connection could lead to an underdose.

The key to designing good connectors and connections is to make component interactions as intuitive as possible and to provide positive feedback. Intuitiveness usually stems from a variety of characteristics, including the following sample:

- **Shape coding.** Shape-coded components normally form a unique pair, precluding connection to any other components on a given product. Shape coding is sometimes called "keying," referring to parts that go together in the same, exclusive manner as a specific key to a specific door lock. This type of coding is illustrated in Figure 13.43.

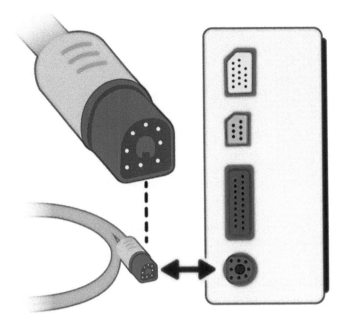

FIGURE 13.43 Saturated oxygen (SP02) cable is shaped (i.e., keyed) and color-coded magenta to ensure attach to the associated port on the patient monitor.

- **Color coding**. Color coding provides a conspicuous sensory cue that specific components of that color go together. Such coding is also illustrated by the patient monitor example presented earlier. Note that the cable itself is colored magenta matched to the color of the connector and port.
- **Tactile feedback**. Feeling a distinct click can tell the user, affirmatively, that they have made a secure connection between two components. The click often comes from the snap-action of temporarily displaced part returning to its starting point. It is important for the feedback to be strong enough to be readily detected in the expected use scenarios. For instance, it might need to be a particularly strong click to be noticeable by an emergency medical technician assembling a device in a moving ambulance.
- **Audible feedback**. Just as tactile feedback can signal a secure connection, so can audible feedback. It is common for a "click," "snap," or equivalent sound to indicate a secure connection and the absence of such a signal to indicate a problem.
- **Labeling.** Labels can guide users to make proper connections. Such labels might include text or icons that indicate the component name or function, as shown in Figure 13.44 depicting oxygen and vacuum plug outlets. Labels can also indicate the order in which the user should make certain connections. For example, two components might be labeled "1," indicating the user should connect these two components together first, followed by the other two components labeled "2."

FIGURE 13.44 These oxygen and vacuum plug outlets are labeled to facilitate easy recognition when making connections.

- **Status indication**. It should be possible to confirm a secure connection by looking at it, or perhaps tugging on the wire/cable/tube because this is how users are inclined to do it. But, there are other ways to indicate the status of a given connection. One is to visually differentiate between components that are securely connected versus not, perhaps by having a colored portion of the connector visible versus hidden within the connector port. Another is to incorporate a sensor that can signal the connection status, perhaps by triggering a change on a display (e.g., appearance of new data, such as a blood pressure waveform and numeric values

FIGURES 13.45 AND 13.46 Three green lights are illuminated (top) when the plane's nose gear and landing gear are fully lowered and locked in place (bottom).

FIGURES 13.45 AND 13.46   (Continued)

appearing in their allotted screen area). Conversely, a device could display a message, such as "Sensor Not Detected" to bring attention to a connection problem. In principle, this approach is conceptually similar to illuminating three green lights in an airplane cockpit when the plane's nose gear and two main landing gear (left and right) are fully lowered and locked in place.

- **Incomplete connection prevention**. There are several ways to prevent an incomplete connection. One is to design components to essentially repel each other, perhaps using spring-action or cam-action, if they are not securely connected. Ostensibly, the goal is to engineer the connection so that it has binary states—securely connected or fully disconnected—and preclude a partial (i.e., loose) connection. Another preventive measure is to stop progression to a subsequent step if a connection has not been made. In other words, incorporate an electronic or some other form of interlock.
- **Placement**. A significant part of risk reduction is making failures and errors readily apparent to users. As such, it helps for users to be able to see connections when visual detection is the only way to tell if there is a poor connection. This is why many medical devices locate connection ports on the front and side panels as opposed to the back panel.

SAMPLE USER INTERFACE REQUIREMENTS FOR CONNECTORS

- Connections shall be shape-coded such that inappropriate connections to other connection ports on the device are physically impossible.

- Matching connector ports and connectors shall be color-coded.
- The system shall provide tactile feedback when the user creates a secure connection between the power port and power plug.
- The system shall provide auditory feedback when the user creates a secure connection between the power port and power plug.
- The power port shall be designed such that the power plug will not stay in the port unless it is securely connected to the port.

COLORS

Color can be a powerful element when used properly to attract attention and convey meaning to people interacting with hardware, software, and documents. Conversely, it also can create visual clutter when applied indiscriminately or for strictly aesthetic purposes. Ultimately, color should be viewed as a precious resource that should not be wasted.

In Chapter 13—Warnings, we discuss the elements of an effective warning, including a colorful header containing a signal word: DANGER on a red background, WARNING on an orange background, and CAUTION on a yellow background. In these cases, color signals the seriousness of a hazard and the consequences of exposure.

Color can also differentiate a product's interactive parts. For example, various interactive parts of the hospital bed shown in Figure 13.47 are colored to draw attention and, to some degree, signal their purpose. In this example, the handle that drops the bed into a flat position and makes the air mattress firm to facilitate CPR is colored red; the footbrakes levers are color-coded orange (brakes applied) and green (brakes released); and handles used to lock and unlock the siderails are color-coded blue.

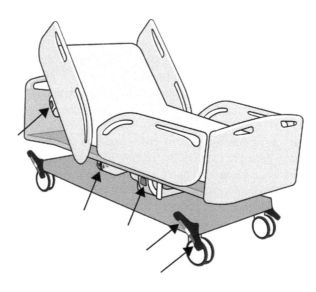

FIGURE 13.47 Hospital bed uses color to differentiate the interactive parts.

Importantly, accommodations should be made for people with various forms of color blindness. For example, it may be problematic to differentiate buttons on a control panel by the colors blue and green and then provide instructions that ask the user to press the green button.

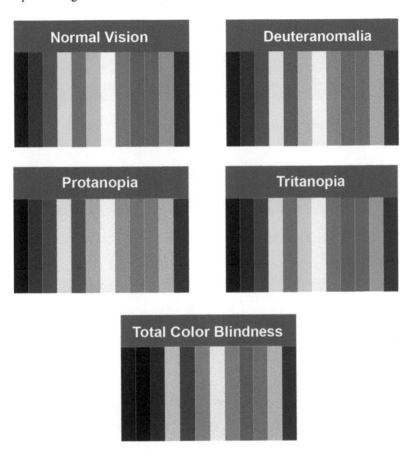

FIGURE 13.48   Appearance of colors based on various types of color blindness.

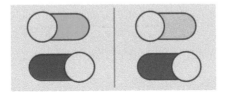

FIGURE 13.49   Green/red slider switches as they would appear to a person with normal vision (left) versus someone who had Deuteranope (red-green color blindness) (right).

It is also important to consider which colors are best for foregrounds and backgrounds to avoid jarring–looking combinations as well as illegible text and numerals.

You can maximize the legibility of labels, for example, by using black text on a white background or white text on a black background. Slightly less extreme pairings work as well, but legibility suffers as foregrounds and backgrounds have less contrast. A classically bad combination is red text on a blue background, as shown in the following figure:

The five boxing wizards jump quickly.

This white text and black background pairing has an excellent contrast ratio of 21:1.

The five boxing wizards jump quickly.

This yellow text and dark gray background pairing has an adequate contrast ratio of 9:1.

The five boxing wizards jump quickly.

This white text and light gray background pairing has a poor contrast ratio of 1.8:1.

The five boxing wizards jump quickly.

This red text and blue background pairing has a poor contrast ratio of 1.4:1.

In many cases, medical devices need to ascribe to standards that help with the identification of contents, purposes, and associated risks. Here is an example of color coding used for medical gas fittings and connections.

Gas	U.S. Color Code		ISO Color Code	
Carbon Dioxide		Gray		Gray
He-O$_2$		Brown Green		Brown White
Instrument Air		Red	N/A	
Medial Air		Yellow		Black White
Nitrogen		Black		Black
Nitrous Oxide		Blue		Blue
O$_2$-He		Green Brown		White Brown
Oxygen		Green		White
Vacuum		White		Yellow
Waste (Evac)		Purple		Purple

FIGURE 13.50   Color coding used for medical gas fittings and connections.

SAMPLE USER INTERFACE REQUIREMENTS FOR COLORS

General
- Colors shall match population conventions established by applicable standards. Example: Colors pertaining to medical gases shall match the color standards presented in ISO 32:1977(en) Gas cylinders for medical use—Marking for identification of content.

- When color coding is used to communicate critical information, there shall be a redundant means of communication (i.e., tactile coding, shape coding, and labeling).

- To the maximum extent possible while also ascribing to international color conventions, colors used to communicate critical information shall be visually distinct from each other to accommodate individuals who have forms of color blindness that can make it *difficult to differentiate* the following colors from each other or not perceive color at all:

  - Red from green
  - Blue from green
  - Yellow from red
  - Purple and red
  - Yellow and pink

- Labels should use alphanumeric characters and background colors that ensure adequate legibility achieved by ensuring a contrast ratio of at least 7:1.

Software

- White text on a medium red background shall signal a failure and indicate a danger.

- Black text on an orange background shall indicate a potential problem; communicate a warning.

- Text on a yellow background shall indicate a need for caution.

- White text on a medium-dark green background shall indicate an active process or activating a process or an action.

- White text on a medium-dark blue background shall indicate a notification.

- Colored buttons shall change color when selected (e.g., become a darker shade when selected [i.e., pressed])

- Colored buttons shall change color when unavailable (e.g., become a lighter shade [i.e., "grayed out" but not necessarily gray] when unavailable).

- Waveforms shall be displayed on a black background.

- Adjustments to color assignments (e.g., color of numeric values) shall be made via a system administrator feature and not available to routine users.

CLEANING

You might think that hospitals and other clinical environments are clean places, considering their purpose. However, this is not necessarily the case. In 2010, the *New York Times* reported on conclusions drawn by the Centers for Disease Control and Prevention that "roughly 1.7 million hospital-associated infections, from all types of bacteria combined, cause or contribute to 99,000 deaths each year."[7] In the ensuing years, healthcare facilities have taken many steps to "clean up their act." Their focus has been not only on ensuring beds, curtains, and floors are properly cleaned but also on getting medical devices clean in view of numerous adverse events.

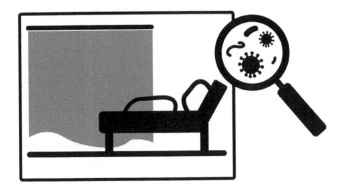

FIGURE 13.51   Hospitals are a hot spot for bacteria.

Clearly, a contaminated medical device is a dangerous medical device. Contamination can lead directly to serious infections among people who very likely are already vulnerable. Therefore, it is important for a medical device to facilitate its cleaning to a high standard.

FIGURES 13.52 AND 13.53   Sample of medical devices used in clinical settings that require cleaning between cases: heart–lung machine (left) and C-arm X-ray machine (right).

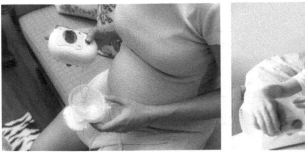

FIGURES 13.54 AND 13.55   Sample of medical devices used in home settings that require daily cleaning: breast pump (left) and nebulizer (right).

Duodenoscopes are one type of medical device that has made headlines due to inadequate cleaning and sterilization leading to patient infections. Clinicians use duodenoscopes to inspect the duodenum via the upper gastrointestinal tract (i.e., passing a scope through the mouth, esophagus, and stomach to reach the short but important gateway segment leading into the small intestine). After use, the scope needs thorough cleaning; a process that can include hundreds of discrete steps and has only been partially automated. The process includes wiping surfaces, flushing internal channels with cleaning solution, and brushing intricate mechanisms. The cleaning task is complex to the point that it induces use errors and is driving some manufacturers to produce scopes with disposable elements that preclude the need for cleaning.

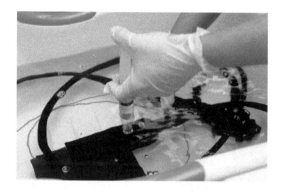

FIGURE 13.56   Flushing an endoscope during manual reprocessing.

Another medical device that has made headlines due to inadequate cleaning is blood heater/cooler devices with ventilation fans that were blowing germs around operating rooms and into open surgical sites.[8] It turns out that such devices were not being adequately cleaned and that the fans were blowing pathogens into the air (and then into the patient). As a result, much attention has been paid to making the devices easier to clean and supporting maintenance personnel with better instructions.

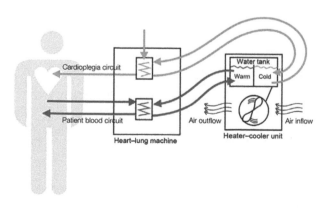

FIGURE 13.57   Heater–cooler circuit during cardiovascular surgery.

In view of the cases described earlier and many more, you can see that medical device cleanability is quite important. Better cleanability comes from the following:

- Hardware that resists contamination and enables rapid and complete cleaning
- Clear instructions
- Effective training

Later, we present sample user interface requirements leading to easy-to-clean hardware. Chapter 13—Instructions provides user requirements pertaining to IFUs (which can include cleaning instructions). Notably, we consider effective training something that also may be specified but that falls outside the scope of this book.

SAMPLE USER INTERFACE REQUIREMENTS FOR CLEANING

Configuration

- External components shall have a minimum number of joints, parting lines, cavities, through holes, crevasses, and similar features that could trap contaminants.
- Components shall be contoured to resist fluid pooling.
- Disassembly for cleaning shall not require the use of any tools.
- Internal components (e.g., fans) shall be accessible for cleaning.

Material and Finish

- On-product label legibility shall not degrade before the 100,000th cleaning procedure.
- Metal surfaces shall be smooth except where texture is required to ensure a secure hand grip or serve another, functionally essential purpose.
- Surfaces subject to contamination shall have finishes that facilitate visual detection of contamination.
  - Example: The device shall have a finish that readily reveals blood contamination (i.e., its exterior color shall have a contrast ratio of 3:1 with dark red [hex #8b0000).

Labeling

- Components requiring a special means of cleaning shall be labeled accordingly.
- Any components that require cleaning but that should not be immersed in fluid shall be labeled accordingly.

Single Use

- Components that are impossible to clean shall be single use (i.e., disposable).
- Single-use components shall be labeled as such using this symbol:[9]

FIGURE 13.58   Single-use symbol.

INSTRUCTIONS

IFUs might be the most disparaged and impoverished part of a medical product's user interface. Some product developers regard IFUs as a necessity—a regulatory mandate—but, otherwise, of limited value. Their common and inaccurate view is that users ignore IFUs. This might come from their own experience trying to use products just out of the box and without looking at the instructions. And, this might work well for a smartphone, noting that there are usually little or no consequences to using the product incorrectly. However, this is not the case with a majority of medical products. On the contrary, use errors can be harmful and sometimes deadly. Accordingly, medical product IFUs are important and warrant a significant investment in producing high-quality content and wrapping it in a good design.

Here are just a few principles for the design of an IFU that will facilitate a medical product's safe, effective, and satisfying use.

- **Cover.** Users are likely to appreciate it when an IFU has a cover that unmistakably indicates the document's purpose and identifies the associated product. This can be readily accomplished by boldly labeling the document with the given product name, the title "Instructions for Use."
- **Table of contents.** Few IFU readers are likely to read the document cover-to-cover. Rather, they are likely to jump around among topics of greatest interest. A table of contents enables quick access to content.
- **Index.** For the same reasons stated earlier regarding a table of contents, a comprehensive index enables quick access to content.
- **Headings and subheadings.** IFU readers will appreciate the liberal use of heading and subheadings as a means to aid their navigation through the document's contents. Users tend to favor hanging headings that appear as topics in the left most position on a page with the associated content indented, thereby enabling them to quickly find their place in the document.
- **Active voice construction.** Writing narrative content in the active voice, as opposed to the passive voice, tends to make a document more engaging to

readers. The active voice can be less wordy, speaking directly to the user. Here is an example of the two writing styles:

- *Passive:* An alarm tone may be temporarily silenced by the user by pressing the Audio Pause button.
- *Active:* Press the Audio Pause button to temporarily silence an alarm tone.
- **Procedural step numbering.** It is beneficial to differentiate procedural steps from other types of content. You can do this readily by numbering the steps. This has the extra benefit of helping users to keep track of where they are in a sequence of steps.
- **Graphics.** Pretty much everyone prefers to read instructions that include lots of graphics/illustrations that either supplement or supplant text. Accordingly, users tend to favor and pay closer attention to IFUs featuring lots of them. However, designers can go overboard with too many illustrations, so it helps to establish rules for their use, such as to make procedure steps clearer and make warnings more attention-getting and communicative.
- **Use of color.** Historically, it has been less expensive to print IFUs in black and white, which enables shades of gray by printing in "halftones." However, IFUs can benefit greatly from the use of color to aid with document navigation, instructional clarity, and establishing a hierarchy of information importance. Therefore, there is a strong argument favoring ideally full-color IFUs over those using just a few colors of sticking to monochrome.
- **Inclusivity.** IFUs should be written and designed to address the user population as broadly as possible in view of its diversity. Accordingly, any photographs might depict people of various sex, gender, skin tone, attire, and other demographic factors. Additionally, graphics can depict skin in an inclusive color, such as the one shown in Figure 13.59.

FIGURE 13.59   One option for depicting skin color in graphics is Hex#C68642.

- **Reading level and vocabulary.** Also related to inclusivity, an IFU should be written to suit the users in terms of reading level and vocabulary. For example, it might be necessary to write an IFU intended for general public use to the eighth-grade reading level and avoid the use of unfamiliar medical terms. In contrast, an IFU intended for use by clinicians might be written at a higher reading level and employ a vocabulary that is familiar to nurses, including common abbreviations and "shorthand" expressions. There are several methods for measuring reading level, such as the Flesch-Kincaid readability formula and the Readability Statistics feature in Microsoft Word. Of course, good comprehension should be evaluated by means of usability testing (including so-called knowledge testing).

- **Language.** A single IFU is sometimes written in two or more languages when the intended user population has a large proportion of individuals who speak different first languages. This avoids the problem of a particular user turning to an IFU for guidance and finding that it is not written in the language with which he/she/they are most familiar. In the United States, for example, IFUs are often written in both English and Spanish. When an IFU is written in multiple languages, it helps for the content to be distinctly separated in some manner, such as reversing the page's top-bottom orientation. In such cases, an IFU has two front covers that sandwich the body content, and no back cover per se.
- **Embedded warnings.** A comprehensive IFU may include a dedicated warnings summary, which may actually include warnings and less severe messages, such as cautions and notices. However, it is important to present the same content in context in the body of an IFU. For example, if the user should disinfect a drug vial's septum before inserting a needle into it, the warning (or perhaps caution) should be presented right along with the procedural guidance on how to draw fluid into the syringe. It helps for warnings to assume a distinct format (i.e., an appearance that is different from surrounding procedural steps), such as the one shown in Figure 13.60.

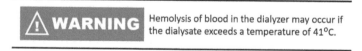

FIGURE 13.60   Warning describing hemolysis of blood.

- **Troubleshooting section.** Users often turn to the IFU for guidance when they run into a problem using a given medical product. Moreover, they look for a quick solution. This is where a dedicated troubleshooting section can be a big help. Such sections usually take the form of a table that includes a column of problems and an adjacent column of solutions.
- **Durability.** Some IFUs live tough lives in their intended use environments. For this reason, it can be helpful to plasticize the cover and pages in some manner, making them less prone to tearing as well as impervious to moisture and fluid.
- **Binding.** It might sound like a minor detail, but spiral or ring binding an IFU with many pages makes it far more usable. The problem with a perfect-bound IFU, for example, is that it is prone to close onto itself if you don't hold it open with a hand or weight. In comparison, a spiral- or ring-bound document can be opened to the page of interest and allow the user to quickly turn to interacting with the given medical product using both hands.

- The user manual shall have a water-resistant, flexible cover.
- "Instructions for Use," or its equivalent, shall be printed on the document's front in ≥32 points.
- The IFU shall be printed on nontransparent, ≥20 lb paper.
- The IFU shall be paginated.
- The IFU shall be printed in full color.
- The IFU shall be spiral or ring bound.
- The IFU shall have a table of contents.
- The IFU shall have an index.
- The IFU shall be written to accommodate users at the fifth-grade reading level, as measured by the Flesch-Kincaid readability formula.
- All major procedural steps shall be numbered.
- All major procedural steps shall be complemented by an informative graphic.
- The overall set of graphics should depict the skin in a consistent color.
- Warnings, cautions, and notices shall be presented in association with procedural steps (i.e., embedded with the pertinent IFU content).

PACKAGES

We could have discarded the topic of medical device packages (pun intended) as unworthy of inclusion in this section of the book, which provides only a sample of use interface design considerations. However, that is not how we look at packages and the role they play in assuring the safe, effective, and satisfying use of medical products.

In fact, the quality of a package can make the difference between a good and bad user experience with a medical product.

- **Means of opening.** Logically, people are unlikely to have much interest in a package, being far more interested in using the contents. Also, they are often in a rush to remove the contents, perhaps because they will serve a time-critical purpose. Accordingly, packages should be designed so that the means of opening them is immediately self-evident and simple. This is where a feature, such as a "Tear here" label or visually distinct pull tabs may be the user's friend. Ideally, opening a package should consume little time, not require special tools, and require only moderate force (i.e., hand strength).

FIGURE 13.61   Sterile packaging often features visually distinct pull tabs.

- **Closure.** Packages that contain many items and will be opened and closed repeatedly should resist damage due to opening. Also, the means of closure should be self-evident, easy to complete, and effective (i.e., the package stays closed). In some cases, the best solution might be a removable and replaceable cover. A hinged cover will keep the cover together with the container.
- **Label.** Users appreciate it when a package's primary contents are clearly labeled. Large, bold text is a good start and a picture or illustration of the contents is also helpful if the package is not see-through. Additionally, the label should be sufficiently distinct that the item cannot be mistaken for a similar but different one. This is particularly important for the package of a combination product, such as an inhaler, that is available in multiple doses and formulations.

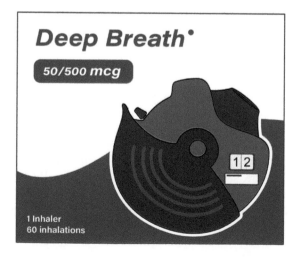

FIGURE 13.62   Packages that include a product illustration are more easily recognized by users.

FIGURE 13.63   Clearly labeled packages are easier to recognize among many packages, such as on a pharmacy shelf.

- **Expiration date.** Using an expired medical product can be a safety hazard. Therefore, expiration dates should be conspicuous and legible. Poor examples take the form of small, alphanumeric character strings, printed in low resolution in an out-of-the-way place on a carton, for example. Better examples take the form of larger, highly legible alphanumeric character strings placed where the user is likely to see them.
- **Relevant instructions.** Users appreciate when the package presents contextually useful instructions. For example, there are some injected medications that must be stored in the refrigerator. Although the IFU includes storage information, users who do not open the package will not find the IFU, and thus will not have the opportunity to see the storage information. These users might then incorrectly store the package at room temperature. A better design would be to print "Store in the refrigerator" on the package, thereby enabling the user to see the instruction it is most relevant.
- **Content retention.** Some medical product packages need to hold several items in place until they are needed, at which point they need to separate readily from whatever is holding them in place. This means that package features, such as snaps (i.e., indentations and undercuts that hold an item in place), need to be "fine-tuned" to have just the right retention capability but not require more than moderate force to disconnect them. Also, when several elements are packaged together in a kit (a packaging approach called "kitting"), care should be paid to creating enough space around the parts (i.e., within the cavities) to grasp them easily and securely.

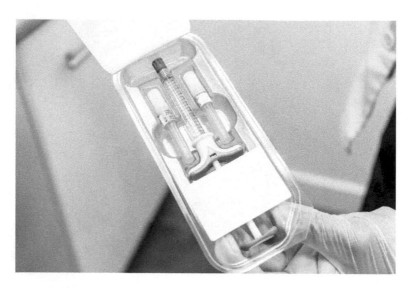

FIGURE 13.64   Tray includes space for users to get their fingers into storage compartments to remove components.

- **Content order (layers).** When a package contains many parts, it is help-ful for the layers to be organized so that they can be removed in the order in which they are needed. As such, a given kit might present components required for the first part of a procedure on top, with the subsequent steps' components on the lower layers.

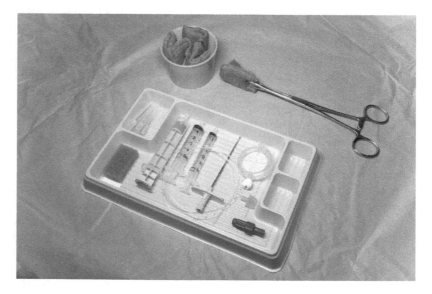

FIGURE 13.65   Epidural catheter overlays the syringes in this epidural kit, matching the order of use (syringes used to deliver medication once catheter is placed).

- **Storage orientation.** Packages should be designed to complement how they will be stored. For example, packages may be stacked like pizza boxes, hung from hooks, tossed into a bin, or placed in a storage pouch. Whenever feasible, labels should be visible when the packages are stored in the intended and common manners.

FIGURE 13.66   Example of hospital storage room.

SAMPLE USER INTERFACE REQUIREMENTS FOR PACKAGING

- The product name shall be printed on the package in ≥24 points.
- The product name shall be printed on two or more package panels.
- The expiration date shall be printed on the package in ≥12 points.
- A product photograph or illustration shall be printed on the package.
- The package shall provide the means to hang it on a hook.
- The package shall provide a visual indication of its means of opening (e.g., "tear here" text).
- The package shall indicate it must be stored in the refrigerator.

NOTES

1   Association for the Advancement of Medical Instrumentation. 2009. *ANSI/AAMI HE75–2009: Human Factors Engineering—Design of Medical Devices*. Arlington, VA: Association for the Advancement of Medical Instrumentation.

2   Gaglani, S. 2012. *Combating Alarm Fatigue: Interview with Philips' Chief Medical Information Officer, Joe Frassica.* Medgadget. www.medgadget.com/2012/10/combating-alarm-fatigue-interview-with-philips-chief-medical-information-officer-joe-frassica.html

3   American National Standards Institute, Inc. 2017. *ANSI Z535.1–2017.* Washington, DC: American National Standards Institute, Inc. https://www.ansi.org/

4   International Organization for Standardization. 2006. *IEC 60601–1–8:2006 Medical Electrical Equipment—Part 1-8: General Requirements for Basic Safety and Essential Performance—Collateral Standard: General Requirements, Tests and Guidance for Alarm Systems in Medical Electrical Equipment and Medical Electrical Systems.* Geneva, Switzerland: International Electrotechnical Commission. https://www.iec.ch/

5   International Organization for Standardization. 2006. *IEC 60601–1–8:2006 Medical Electrical Equipment—Part 1-8: General Requirements for Basic Safety and Essential Performance—Collateral Standard: General Requirements, Tests and Guidance for Alarm Systems in Medical Electrical Equipment and Medical Electrical Systems.* Geneva, Switzerland: International Organization for Standardization. www.iso.org

6   American National Standards Institute, Inc. 2007. *ANSI Z535.4:2007: Product Safety Signs and Labels.* Washington, DC: American National Standards Institute, Inc. https://www.ansi.org/

7   Pollack, A. 2010. Rising Threat of Infections Unfazed by Antibiotics. *The New York Times.* New York. www.nytimes.com/2010/02/27/business/27germ.html?em=&adxnnl=1&adxnnlx=1267412412-yP2bfl/3pu4+g34XVmluJA&_r=0

8   Sommerstein, R., Rüegg, C., Kohler, P., et al. 2016. Transmission of Mycobacterium Chimaera from Heater-Cooler Units during Cardiac Surgery Despite an Ultraclean Air Ventilation System. *Emerging Infectious Diseases* 22.6: 1008.

9   Based on the symbol presented in: International Organization for Standardization. 2016. *15223–1:2016 Medical Devices—Symbols to be Used with Medical Device Labels, Labelling and Information to be Supplied—Part 1: General Requirements.* Geneva, Switzerland: International Organization for Standardization.www.iso.org

14 Expanded Example for a Glucose Meter

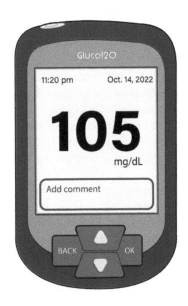

ABOUT THE EXPANDED EXAMPLE

- The user interface requirements in this expanded example are for a blood glucose meter. The requirements address the glucose meter as a standalone component, thereby covering the meter, associated user documentation, and storage case. Developers of such devices might choose to extend the requirements to cover such elements as lancing device, lancets, test solution, and logbook, depending on what they are developing in-house versus sourcing from other manufacturers with approved or cleared products. For the purposes of this example, we focused on the meter, associated user documentation, and storage case, and assumed the other system components would be addressed separately.
- We begin the example with a product description to give readers a sense for a blood glucose meter's function, users, and use environment.
- The user interface requirements are hypothetical but inspired in many cases by user-friendly design features that we have found or read about in actual products.
- We followed a customized outline that has elements of the possible organizing schemes presented in Chapter 5.

DOI: 10.1201/9781003029717-14

- We present over 200 user interface requirements for the blood glucose meter. That said, the example is still not comprehensive. We imagine that the number of requirements could increase by perhaps 50% as a result of a research-driven development effort.
- We refrained from writing user interface requirements that are overly progressive. However, you'll see that, writing them in accordance with earlier advice to prescribe desirable interactive characteristics, but not design solutions per se (Chapters 4 and 5), leaves the door open to innovation.
- The example includes many specific, verifiable requirements. That said, we also present some general or high-level requirements that are not sufficiently specific to be verified as is. You can think of each general requirement as a precursor to several more specific requirements that would arise to achieve the goal conveyed by the general requirement. We included a handful of these general requirements to give readers a sense for the design objective driven by a user need, which is not always immediately obvious when reading highly specific requirements. If this was a real use specification, we would discard the general requirements in favor of ones that are verifiable.
- The user interface requirements are intended to be illustrative rather than definitive for the blood glucose meter. If we were preparing user interface requirements for an actual product, we would have conducted extensive user research, possibly complemented by other activities described in this book, to arrive at a more definitive set. Therefore, we present the example for educational purposes only. In this regard, some of the requirements might be off-the-mark from those that might have resulted from the aforementioned research effort.
- We conclude the example by illustrating a few, possible design solutions that fulfill specific requirements. In many cases, the illustrated solution is just one of many possible ways to satisfy the associated user interface requirement.

ABOUT GLUCOSE METERS

To contextualize the following user interface requirements, we will explain the purpose of a glucose meter and how they usually work.

PURPOSE

A glucose meter is a device that enables users to determine the sugar (i.e., glucose) content of their blood. This is an important task that people with diabetes mellitus can perform to help manage their disease. If they perform a blood test and determine that their glucose level is high, they may eat less of certain foods, particularly sugary items and carbohydrates, exercise, or perhaps take medication (e.g., insulin). If they determine that their glucose level is low, they may eat more of certain foods, particularly sugary items and carbohydrates, take glucose supplements, reduce their exercise, or perhaps take less medication (e.g., inject less insulin).

USER INTERACTION

Glucose meters are typically handheld devices that are similar in size (height and width) to a credit card. The meters and accessories are usually stored in a handy case. Performing a blood test requires the user to insert a device-specific test strip into the meter, prick their finger with a small blade, apply a small droplet of blood to the strip, and then wait for a few seconds for the meter to display a test result. The user often has the option to annotate the test result with details, such as if they just had a meal or exercised, and store that information in the meter's memory. The process usually wraps up with the user disposing of the used test strip and putting the meter and accessories back in the storage case.

SAMPLE GLUCOSE METERS AND KITS

In Figures 14.1–14.4, we show some sample glucose meters and kits. These are just a few of the many products available on the market in various regions of the world.

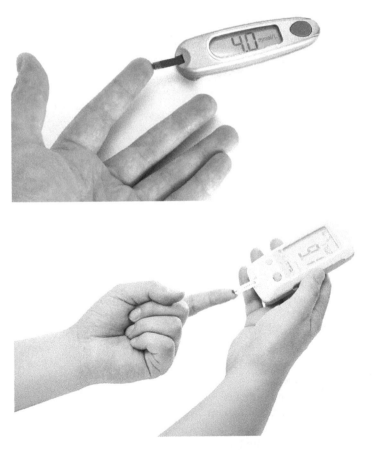

FIGURES 14.1–14.4   Sample blood glucose meters available on the market.

FIGURES 14.1–14.4   (Continued)

USER INTERFACE REQUIREMENTS

The sample user interface requirements include the following categories:

- Hardware
- Software
- Voice guidance
- Power management
- Text, numerals, and symbols
- Warnings, cautions, and notices
- Annotations, analytics, and advisories
- Errors, malfunctions, and quality checks
- Data storage

- User manual
- Quick reference guide (QRG)

HARDWARE

Physical Device

1. The meter shall weigh between 0.2 and 1 lb.[1]
2. The meter shall withstand a drop onto a hard surface (e.g., concrete) from 1 meter without sustaining damage.[2]
3. The meter shall have no sharp points or edges that could potentially tear an exam glove or lacerate the user's skin.
4. The meter shall have no pinch points.
5. The meter shall rest (i.e., sit with screen facing up) stably on a flat surface (e.g., table).
6. The meter shall have a nonslip texture (e.g., material with a high coefficient of friction).
7. The meter's finish shall readily reveal blood contamination (i.e., its exterior color shall have a contrast ratio of 3:1 with dark red [hex #8b0000]).
8. The meter shall be water-resistant up to 3 feet for up to 30 minutes.

Display

9. The meter shall have a 24-bit color display.
10. The display shall be backlighted.
11. The backlight shall illuminate when the device is powered on.
12. The meter's screen shall remain impervious to scratches at point 5 on Mohs scale of mineral hardness during a scratch resistance test.
13. The meter shall withstand 350 Newtons of pressure during a Screen Toughness test.
14. The meter's display shall have an antiglare treatment (e.g., light antiglare coating with 25% haze value).
15. The meter's display shall be invulnerable to internal fogging.
16. The meter's display shall be clear and remain so (e.g., no yellowing) over the course of three years (the expected life of the meter).
17. The screen pixels shall have an aspect ratio of 1:1.
18. The display shall be pixel addressable (i.e., not a segmented character display).
19. The display resolution shall be ≥100 dpi.
20. The display shall have an adjustable luminance of up to 750 cd/m^2 at a 2-feet viewing distance.

Test Strip Insertion Port

21. The test strip port shall be sized to fit compatible test strips (strips 0.5–0.8 mm wide).

22. The test strip port shall be appropriately deep, such that a ≥1.3 cm portion of the test strip remains external to the meter when a strip is fully inserted.

23. The test strip port shall be labeled.

24. The test strip port should include a label indicating the correct test strip orientation for insertion.

25. The test strip port should be shaped to guide the test strip into it.

26. The test strip port shall be visually differentiated from other parts of the meter's body (e.g., a different color).

27. The test strip port shall illuminate when the meter is on but the user has not yet inserted a test strip.

28. The test strip port shall remain illuminated when the user inserts the test strip and before the glucose level has been displayed.

29. The test strip port's light shall flash (60–120 Hz) when the test is done and the used strip may be removed.

30. The meter shall provide tactile feedback when the user fully inserts the test strip.

31. The meter shall enable the user to eject a used test strip without having to touch the test strip.

32. The meter's test strip removal force shall be 1.0–2.0 lbF so that the meter retains the used test strip until the user deliberately removes it.

Battery

33. The dormant battery (i.e., in the device, in its packaging, and before sale and use) shall be protected from dissipating.

34. A fully charged battery shall enable at least 2,000 minutes of use.

35. The meter should indicate the need to change the battery(ies) when the battery is at 10% of its full charge.

36. The meter backlight shall no longer illuminate once the battery is an estimated 97.5% discharged.

37. The meter shall enable the user to replace the battery(ies) without using tools (e.g., screwdriver).

38. The battery compartment shall be labeled.

39. The battery compartment shall be protected against accidental opening.

40. The battery compartment shall be protected against partial closure that makes it susceptible to inadvertent opening.

41. It shall be possible to open and close the battery compartment using one hand.

42. Changing the battery(ies) shall not cause any data loss.

Storage Case

43. The meter shall be supplied with a storage case.

44. The storage case open/close mechanism (e.g., zipper, flap with a snap) should be visually distinguished from the rest of the case (e.g., different color).

45. The case shall incorporate some means of protecting the meter from moderate blunt forces and crushing.

46. At a minimum, the storage case shall hold the meter, user manual, QRG, and necessary accessories (i.e., test solution, test strips, lancet device, and lancets).

Labeling

47. The meter shall be labeled with the device's name.

48. The meter shall be labeled with the device's serial number.

49. The meter shall be labeled with the manufacturer's name.

50. The meter shall be labeled with the software version.

51. The meter shall use FDA-approved, standard symbols where applicable.[3]

52. Button labels shall be placed on the button faces or just above the buttons.

53. During the life of the product, labels shall resist aging due to humidity exposure that would reduce their legibility.

54. During the life of the product, labels shall resist aging due to sunlight (specifically UV light) exposure that would reduce their legibility.

55. During the life of the product, labels shall resist wear, such as due to repeated finger touches, that would reduce their legibility.

Other

56. The meter shall have an optional lanyard.

57. The meter shall remain fully functional in ambient temperature in the range of 40°F—120°F (4.44°C—48.89°C).

58. The meter shall remain fully functional in humidity in the range of 10%–90%.

Buttons (Keys)

59. The meter shall have ≤6 dedicated buttons.

60. Buttons shall be placed only on the front of the meter.

61. Buttons shall be either flat or concave, but not convex.

62. Buttons shall be shape coded to differentiate their purposes.

63. Buttons shall be ≥0.5 inches (1.27 cm) across in all directions, regardless of shape.

64. Button centers shall be spaced ≥0.75 inches (1.9 cm) apart.

65. Buttons shall have debounce protection.

66. Each button shall perform a single purpose (e.g., confirm a setting, scroll on a displayed menu).

67. The meter shall have a dedicated power on/off button.

68. The meter shall have a dedicated "select" (or "OK") button.

69. The meter shall be turned off manually by pressing and holding the power-off button for ≥5 seconds.

70. The meter shall provide the user with a hardware-based means to turn off the backlight.

71. Buttons shall provide tactile feedback (e.g., via button travel or vibration) to confirm that a button press has been detected.

72. Buttons shall provide audible feedback (which may be produced by the button itself or a separate speaker) to confirm that a button press has been detected when the meter's sound is turned on.

73. The meter shall emit an audible sound (≥60 dB) when the user presses and holds any key for the purpose of moving through a data set or options (i.e., when scrolling).

74. Buttons shall provide visual feedback (which may be presented on the display) to confirm that a button press has been detected.

75. Button labels shall contrast with their backgrounds (contrast ratio shall be ≥7:1).

76. Button label alphanumeric characters shall subtend a visual angle of ≥24 arc minutes. Given a potential viewing distance of up to 24 inches—which equates to holding the meter at a long arm's distance—the smallest numbers and characters shall be ≥0.17 inches (≥12 points).[4]

Speaker (Audio Emitter)

77. Emitted sounds shall be adjustable over the range of 0–70 dB.

78. Emitted sounds shall be in the range of 400–1500 Hz.[5]

79. The audio output device shall be protected against being fully covered by a separate object (e.g., hand, table surface, pillow).

80. The meter shall emit a power-on sound (e.g., an "intro melody").

81. The meter shall emit a power-off sound (e.g., an "outro melody").

82. The meter shall emit an audible alert when the blood test is done/the result is displayed.

83. The meter shall emit an audible alert when it displays a warning.

84. The meter shall emit an audible alert when it displays a caution.

85. The meter shall emit an audible alert when it displays a notice.

86. Audible alerts for warnings, cautions, and notices shall be audibly distinguished from each other (e.g., different sounds).

87. Audible alerts for warnings, cautions, and notices shall be audibly distinguished from other sounds emitted by the meter (e.g., power on/off sounds).

88. The meter shall enable the user to turn off sounds.

SOFTWARE

Setup

89. The meter shall enable the user to clear (i.e., delete) all stored data and return it to factory preset condition.

90. Clearing all data shall require at least four user steps to ensure this does not occur accidentally.

91. The meter shall address the user by first name.

92. The meter shall enable users to set the date, selecting among the following date formats:

   - Day-Month-Year with leading zeros (17/02/2009)
   - Year-Month-Day with leading zeros (2009/02/17)
   - Month name-Day-Year with no leading zeros (February 17, 2009)
   - Month-Day-Year with no leading zeros (2/17/2009)

93. The meter shall enable users to select between time formats: (1) 24-hour/railway/continental/military format, wherein mid-afternoon is displayed as 15:00, and (2) 12-hour format, wherein mid-afternoon is displayed as 3:00 PM.

94. When the meter displays time in the 12-hour format (e.g., 3:15 PM), "PM" should be the same size as the time numerals.

95. The meter shall require the user to confirm the date and time settings.

96. The meter shall enable users to set the glucose units to be mmol/L (millimoles per liter) or mg/dL (milligrams per deciliter).

97. The meter shall enable the user to select among the following languages that are spoken in the targeted markets: Bengali, Mandarin Chinese, Marathi, English, French, German, Hebrew, Hindi, Indonesian, Japanese, Portuguese, Russian, Spanish, Standard Arabic, Swahili, Tamil, Telugu, Turkish, Urdu, Western Punjabi, Wu Chinese

Blood Testing

98. When powered on, the meter should display the last glucose reading for up to 15 seconds.

99. After displaying the most recent glucose reading upon powering on, the meter should direct the user to insert a test strip.

100. The user should be able to access a diagram indicating their last five test sites with one button press from the main screen.

101. The meter should remind the user to wash and dry their hands before lancing (i.e., pricking) their finger to produce a blood droplet.

102. After the user inserts a test strip, the meter shall indicate that the user should (1) lance their finger to produce a blood droplet and (2) apply blood to the inserted test strip.

103. The meter shall not accept a used test strip.

104. If a user inserts a used test strip, the meter shall instruct the user to remove the used test strip and insert a new one.

105. The meter should advise the user about how much blood to apply to the test strip.

106. The meter should indicate where on the test strip to apply blood.

107. The meter shall indicate when the meter has detected blood on the test strip and the test has commenced.

108. The meter should provide a countdown from the time it recognizes that a blood sample has been placed on the test strip and when it will display the result.

109. The meter should produce a test result ≤5 seconds after blood is properly applied to the test strip.

110. The meter shall enable the user to delete the most recent glucose reading if it is judged to be incorrect.

111. The meter shall display a blood test result continuously until the user initiates another task or the meter automatically powers off.

112. The test result shall be displayed with test's associated time and date (i.e., time stamped).

113. The blood glucose reading units of measure shall be displayed on the right side or below the glucose reading.

Smartphone Connectivity

114. The meter shall accept a download of configuration and test data from another meter via an authorized, paired smartphone.

115. The meter shall require the user to enter a code, provided to the user via a secure means, to enable it to be paired to a designated smartphone.

116. The meter shall have the option to be programmed by means of downloading a setup configuration from a smartphone running an associated setup application.

117. The meter shall be configurable without downloading a setup configuration from a smartphone.

118. The meter shall enable users to pair the device with a smartphone to enable data sharing.

119. The meter shall indicate when its smartphone communication capability (e.g., Bluetooth) is turned on.

120. The meter shall indicate when its smartphone communication capability is turned off.

121. The meter shall produce a unique error when the connected phone's date and time settings and the meter's date and time settings do not match.

Voice Guidance

122. The meter shall provide voice guidance pertaining to all tasks.

123. The meter shall enable the user to select among levels of guidance: high, medium, and low.

124. The meter shall enable the user to select a male or female voice.

Power Management

125. The meter's display shall activate ≤1 second after the user presses the power-on control.

126. The meter's display shall activate ≤1 second after the user inserts the test strip.

127. The meter shall automatically power off after 1–2 minutes of inactivity.

128. When the meter is powering off, the display shall provide a countdown from 5 to 0 seconds.

129. Halting the power-off countdown shall require a single action.

130. The meter shall enable the user to select an interval that starts with the most recent user interaction or display change (e.g., displays blood test result) and when it automatically powers off (e.g., 15 seconds, 30 seconds, 45 seconds, 60 seconds).

Text, Numerals, and Symbols

131. On-screen text fonts shall be sans serif (i.e., the letterforms will not have extended features/flourishes or stroke width variations).

132. On-screen alphanumeric characters shall have a "normal" stroke width of 1/12 to 1/6 of the character height.

133. The glucose reading shall be ≥60 points.

134. The glucose reading text shall be black against a white background.

135. The units of measure text shall be black against a white background.

136. The clock shall display hours and minutes.

137. The colon (:) between the hours and minutes shall flash once per second.

Warnings, Cautions, and Notices

138. On-screen warnings shall be presented as white text on a medium-red background.

139. On-screen cautions shall be presented as black text on a medium-orange background.

140. On-screen notices shall be presented as white text on a medium-blue background.

141. The meter shall indicate the battery level in ≥5 increments between fully charged and virtually discharged.

ANNOTATIONS, ANALYTICS, AND ADVISORIES

142. The meter shall enable users to annotate a glucose reading to indicate the test site by selecting among these options:

    • Each finger on each hand
    • Each palm
    • Each forearm

143. The meter shall enable users to annotate a glucose reading to indicate that they recently ate a meal.

144. The meter shall enable users to annotate a glucose reading to indicate that they recently took medication.

145. The meter shall enable the user to annotate a glucose reading to indicate that they recently exercised.

146. The meter shall enable the user to edit preset low and high limits (i.e., establish a target range).

147. The meter shall indicate if each glucose reading is within the target range, above the high limit, or below the low limit.

148. If the blood glucose reading is out of the normal range, the meter shall advise the user to consult a medical professional.

149. If the blood glucose reading is far out of the normal range (above 240 mg/ dL), it shall advise the user to perform another test and, if the result is still far out of the normal range, to consult a medical professional.

150. The meter shall alert the user to a pattern of increasing or decreasing glucose (e.g., a trend of two or more consecutive readings above or below the normal range).

151. The meter shall enable the user to view their average glucose level for the past 1, 7, and 14 days.

152. The meter shall enable the user to view a table of date-stamped, glucose readings for the past 1, 7, and 14 days.

153. The meter shall enable the user to view a graph of glucose readings for the past 1, 7, and 14 days.

154. The meter shall enable the user to view a graph of glucose readings for the past 1, 3, 6, and 12 months.

155. The meter shall enable the user to view a table of date-stamped, glucose readings for each comment type (e.g., exercise, meal, medication) from the past 1, 3, 6, and 12 months.

156. The meter shall enable the user to view a table of date-stamped, glucose readings that are above their normal range from the past 1, 3, 6, and 12 months.

157. The meter shall enable the user to view a table of date-stamped, glucose readings that are below their normal range from the past 1, 3, 6, and 12 months.

ERRORS, MALFUNCTIONS, AND QUALITY CHECKS

158. The meter shall indicate when it has performed a successful self-check during the power-on process.

159. The self-check should be completed within ≤1 second.

160. If the self-check must take longer than 1 second, it shall be completed while the meter displays the last glucose reading.

161. The self-check indication shall be visually diminished (e.g., smaller) compared to higher priority information (i.e., most recent glucose reading and its date and time).

162. The meter shall indicate when it is too cold to perform an accurate glucose test, concurrently indicating its proper operating range.

163. The meter shall indicate when it is too hot to perform an accurate glucose test, concurrently indicating its proper operating range.

164. The meter shall indicate when it has malfunctioned.

165. All error messages shall be simply worded (i.e., include 8 or fewer words).

166. Error messages shall not present only error codes.

167. The meter shall enable the user to perform a test on a known fluid to determine the glucose reading accuracy (i.e., perform a quality check).

168. The meter shall automatically distinguish a quality check fluid from a blood droplet and designate the glucose reading as a quality check.

DATA STORAGE

169. The meter shall store test result data (reading, time, date, annotations) for ≥1 year.

170. When test data memory is full, new test data shall displace the oldest test data, one at a time.

171. The meter shall enable the user to delete test data over a selected interval of time.

172. The meter shall require the user to confirm any action that deletes data, including what might be deemed an incorrect test result.

173. The meter shall enable the user to scan through prior test results in reverse chronological order from the current date and time.

174. The meter shall enable the user to scan through prior test results by specifying a starting point and then moving forward or backward in time.

175. The meter shall enable the user to set reminders to perform a blood test.

176. The meter shall enable the user to set the type of blood test reminder. The options shall not be mutually exclusive and could include (1) audible, (2) visual, (3) tactile, and (4) delivered via a smartphone alert.

USER MANUAL

177. The meter shall be supplied with a physical (i.e., printed) user manual.

178. The title "User Manual," or its equivalent, shall be printed on the document's front in ≥32 points.

179. The user manual shall be printed on nontransparent, ≥20 lb paper.

180. The user manual shall have a water-resistant, flexible cover.

181. The user manual shall be printed in full color.

182. The user manual shall be spiral bound.

183. The user manual shall be written to accommodate users at the eighth-grade reading level, as measured by the Flesch-Kincaid readability formula.

184. The user manual shall be paginated.

185. All major procedural steps shall be numbered.

186. The user manual shall have a table of contents.

187. The user manual shall have an index.

188. All major procedural steps shall be complemented by an informative graphic.

189. If the user manual includes photographs (e.g., on the cover), the photographs shall include individuals of varying racial background, age, gender, and/or body type.

190. The overall set of graphics should depict the skin in an inclusive color (e.g., Hex#C68642).

191. The overall set of graphics should depict the skin in a consistent color.

192. Warnings, cautions, and notices shall be presented together in a summary.

193. Warnings, cautions, and notices shall be presented in association with detailed guidance (i.e., embedded with the pertinent user manual content).

194. The user manual shall have a hierarchical content organization scheme that is reinforced with headings and subheadings.

195. The user manual shall have informative headers and/or footers.

196. All figures and tables shall be titled.

197. Graphics adjacent to explanative text should not be labeled.

198. Graphics that are not directly associated with and described by adjacent text shall be captioned.

199. The user manual shall include definitions of all symbols appearing on the device and software screens.

200. The user manual shall provide instructions about how to clean the meter.

201. The user manual shall include a troubleshooting section.

202. The user manual shall include a helpline phone number.

203. The user manual shall include URL to access a website that provides helpful information.

204. The user manual shall be available online.

205. The user manual shall be available in a mobile version (i.e., viewable on a smartphone).

QUICK REFERENCE GUIDE (QRG)

206. The meter shall be supplied with a QRG.

207. The QRG shall be printed on a waterproof, flexible, and tear-resistant material.

208. The QRG shall fit in the storage case.

209. The title "Quick Reference Guide," or its equivalent, shall be printed on the document in ≥24 points.

210. The QRG should be ≤24 inches (60.96 cm) in width and height.

211. If the QRG is two-sided, the sides (e.g., Side 1, Side 2) shall be labeled.

212. QRG fold lines (if there are any) shall not intersect with text (i.e., break up a word or sentence).

213. The QRG shall be printed in full color.

214. The QRG shall address the needs of a first-time user, guiding them through activating and setting up the meter, as well as performing their first blood test.

215. The QRG shall visually distinguish primary information from secondary information.

216. Warnings, cautions, and notices shall be presented in association with detailed guidance (i.e., embedded with the pertinent QRG content).

217. The QRG shall include a helpline phone number.

218. The QRG shall include URL to access a website that provide helpful information.

219. The QRG shall be available online.

220. The QRG shall be available in a mobile version (i.e., viewable on a smartphone).

Sample Design Solutions That Fulfill User Interface Requirements

Requirement 71

The pushbuttons provide tactile feedback, moving noticeably in response to presses and producing a definitive "click" feel when actuated.

Requirement 133

The blood glucose reading is considerably larger (10.8 mm tall numerals) than any other information on the screen.

Requirements 143 and 145

The device provides the option to indicate whether the user has recently eaten a meal or exercised.

NOTES

1   Zingale, C., Ahlstrom, V., and Kudrick, B. 2005. *Human Factors Guidance for the Use of Handheld, Portable, and Wearable Computing Devices.* U.S. Department of Transportation and Federal Aviation Administration. https://hf.tc.faa.gov/publications/2005-human-factors-guidance-for-the-use-of-handheld/full_text.pdf

2   Damage is defined as a major loss in functionality (e.g., problems powering on, capturing data, or data entry).

3   The FDA recognizes symbols described in: International Organization for Standardization. 2016. *15223–1:2016 Medical Devices—Symbols to be Used with Medical Device Labels, Labelling and Information to be Supplied—Part 1: General Requirements.* Geneva, Switzerland: International Organization for Standardization. www.iso.org

4   Association for the Advancement of Medical Instrumentation. 2009. *ANSI/AAMI HE75–2009: Human Factors Engineering—Design of Medical Devices.* Arlington, VA: Association for the Advancement of Medical Instrumentation.

5   Association for the Advancement of Medical Instrumentation. 2009. *ANSI/AAMI HE75–2009: Human Factors Engineering—Design of Medical Devices.* Arlington, VA: Association for the Advancement of Medical Instrumentation.

Index

9 780367 457471